Innovations in Science Education and Technology

Volume 26

Series editor
Karen C. Cohen
Weston, MA, USA

As technology rapidly matures and impacts on our ability to understand science as well as on the process of science education, this series focuses on in-depth treatment of topics related to our common goal: global improvement in science education. Each research-based book is written by and for researchers, faculty, teachers, students, and educational technologists. Diverse in content and scope, they reflect the increasingly interdisciplinary and multidisciplinary approaches required to effect change and improvement in teaching, policy, and practice and provide an understanding of the use and role of the technologies in bringing benefit globally to all. Book proposals for this series may be submitted to the Publishing Editor: Claudia Acuna E-mail: Claudia.Acuna@springer.com

More information about this series at https://link.springer.com/bookseries/6150

Jesper Haglund • Fredrik Jeppsson
Konrad J. Schönborn
Editors

Thermal Cameras in Science Education

Springer

Editors
Jesper Haglund
Karlstad University
Karlstad, Sweden

Fredrik Jeppsson
Linköping University
Norrköping, Sweden

Konrad J. Schönborn
Linköping University
Norrköping, Sweden

ISSN 1873-1058 ISSN 2213-2236 (electronic)
Innovations in Science Education and Technology
ISBN 978-3-030-85287-0 ISBN 978-3-030-85288-7 (eBook)
https://doi.org/10.1007/978-3-030-85288-7

This Springer imprint is published by the registered company Springer Nature Switzerland AG
The registered company address is: Gewerbestrasse 11, 6330 Cham, Switzerland

Foreword

There is a long history of using sensors to promote inquiry in science education, pioneered notably by Dr. Robert F. Tinker more than three decades ago. In most cases, a few sensors are used to measure certain physical or chemical properties. These sensors are then connected to a computer that logs their measurements. Students typically use a line graph to plot the incoming data as a function of time to visualize the scientific effect under investigation. Through observing how a property changes graphically in response to a stimulus in the real world, students develop a concrete idea about the causal relationship between the variables in question.

Since students can arbitrarily position a sensor in an experiment, educators often need to explicitly prescribe where and when students should put it in order to capture the data as intended. While this type of cookbook-like instruction ensures that students do not miss the chance to observe the due phenomenon, it may deprive them of the opportunities to make their own scientific discoveries and the excitement experienced in such intellectual processes.

Thermal imaging provides a novel technology to support authentic science inquiry without taking away the joy of discovery. A thermal camera automatically gathers a large array of radiometric data for immediately rendering an intuitive, salient heat map visualization of a phenomenon. Such a real-time, full-field visualization of an experiment enables students to discover important details of the emergent phenomena that would otherwise go unnoticed (e.g., the thermal energy released as a result of water molecules making hydrogen bonds with cellulose molecules when a water drop falls onto a piece of dry paper). Data collection is also made as convenient as taking a picture or recording a video just like using a conventional digital camera, freeing students from tedious procedures of scanning an area with a pointwise sensor and focusing them on the fun part of doing science. Such transformative potential would not have been possible without the power of many: Compared with a single thermometer that outputs only a data point at a time, a thermal camera bundles thousands of microbolometers in a small optoelectronic device to produce a large quantity of data at once.

As this book demonstrates, science educators have been tapping into the learning and teaching potential of thermal imaging not long after it was invented. With the plummet of price and the integration with smartphones in recent years, thermal cameras have never been more accessible to students and teachers. However, for the technology to take root and its application to grow in science education, we will need further research and development to convince schools to seriously invest in it. To this end, there may be three directions that are worth considering.

First, it is important to keep informing teachers that thermal imaging can be used to visualize any physical, chemical, and biological processes that absorb or release heat. In other words, *anything that leaves a trace of heat leaves a trace of itself under a thermal camera*. Based on this principle, we can use thermal energy as a universal indicator (passive or active) to investigate many questions in science. For example, what affects the rate of a chemical reaction can be studied using thermal imaging as a qualitative method. In the case of an endothermic reaction such as baking soda with vinegar, a petri dish filled with an aqueous reactant that is initially warm will exhibit a larger drop in temperature than another one that is initially cool when the reaction completes, suggesting that increasing temperature can speed up the reaction. Often, the challenge in broadening the application scope of thermal imaging is that it requires the participation of domain scientists to design, test, and optimize the thermal imaging equivalents of existing experimental techniques in their curricula and determine whether there are sufficient advantages to justify the conversion. This is by no means a trivial task. But if most teachers only view a thermal camera as a tool for seeing heat and are unaware of its broader applications, it would remain unlikely that schools adopt it as a main instrument for their science labs—especially in underserved districts that have a cash-strapped science budget.

Second, it is important to develop different types of apps based on the smartphone versions of thermal cameras to meet diverse needs in formal and informal science education. As a *mobile lab-on-a-chip* that can be used anywhere and connected to the cloud as part of the emerging Internet of Things, a thermal camera has tremendous potential for learning and teaching in different settings, particularly when integrated with other technologies. For example, the Infrared Street View project funded by the U.S. National Science Foundation at the Institute for Future Intelligence is a citizen science program that engages students to create thermal panoramas of their communities to survey their thermal landscapes. The project uses smartphone sensors to determine the location and orientation of a thermal image being captured by a thermal camera attached to the smartphone. Such thermal landscapes provide rich information about building energy efficiency, urban heat islands, solar energy potential, wildfire prevention, and so on that are relevant to personal issues in everyday life and global issues about climate change.

Third, it is important to develop learning and teaching theories to uncover the instructional power of thermal imaging as an example of *augmented cognition*, on a par with the active educational research on augmented reality. Many science concepts are difficult to students because they cannot be readily seen in the real world. One can only imagine how much thermal vision could help students overcome those cognitive barriers by simply empowering them with an additional sense to see a whole new world that is previously hidden from them. On the other hand, we all know that no technology can miraculously become a good educational tool without research to unpack how people may learn and teach with it. Sound theoretical framing and solid empirical evidence based on appropriate principles and approaches from the learning sciences would help us make a more compelling case for thermal imaging to science educators and policy makers.

Without a doubt, the efforts of the small international community represented by the contributors to this book are highly commendable. With their foundational work and continuous commitment, the power of thermal imaging is bound to enlighten more students around the world in the future. It is also our hope that this book will inspire even more applications to push the envelope of this fascinating field.

Chief Scientist
Institute for Future Intelligence
February, 2021

Charles Xie

Contents

List of Contributors

Niclas Åhman Department of Physics and Electrical Engineering, Linnaeus University, Växjö, Sweden

Eugenia Etkina Graduate School of Education, Rutgers, the State university of New Jersey, New Brunswick, NJ, USA

Sofia Folestam Lillerudsgymnasiet, Vålberg, Sweden

Larissa Greinert Ricarda-Huch-Schule Hannover, Hannover, Germany

Jesper Haglund Department of Engineering and Physics, Karlstad University, Karlstad, Sweden

Fredrik Jeppsson Department of Behavioural Sciences and Learning, Linköping University, Norrköping, Sweden

Kalle Juuti Department of Education, University of Helsinki, Helsinki, Finland

Jari Lavonen Department of Education, University of Helsinki, Helsinki, Finland

Anni Loukomies Viikki Teacher Training School, University of Helsinki, Helsinki, Finland

Taina Makkonen Viikki Teacher Training School, University of Helsinki, Helsinki, Finland

Urška Nered Master student at the Faculty for Mathematics and Physics, University of Ljubljana, Ljubljana, Slovenia

Jeffrey Nordine Physics Education, Leibniz Institute for Science and Mathematics Education (IPN), Kiel, Germany

Stefano Oss Department of Physics, University of Trento, Povo, Italy

Karljohan Lundin Palmerius Department of Science and Technology, Linköping University, Norrköping, Sweden

Gorazd Planinšič Faculty for Mathematics and Physics, University of Ljubljana, Ljubljana, Slovenia

Christopher Robin Samuelsson Department of Physics and Astronomy, Uppsala University, Uppsala, Sweden

Konrad J. Schönborn Department of Science and Technology, Linköping University, Norrköping, Sweden

R. Subramaniam National Institute of Education, Nanyang Technological University, Singapore, Singapore

Michael Vollmer Department of Engineering, University of Applied Sciences Brandenburg, Brandenburg an der Havel, Germany

Susanne Weßnigk QUEST-Leibniz-Forschungsschule, Leibniz University Hannover, Hannover, Germany

Anna Westlund Lillerudsgymnasiet, Vålberg, Sweden

Choun Pei Wong National Institute of Education, Nanyang Technological University, Singapore, Singapore

Charles Xie Institute for Future Intelligence, Natick, MA, USA

Chapter 1
Introduction

Jesper Haglund, Fredrik Jeppsson, and Konrad J. Schönborn

Studying thermal phenomena involves exploring physical processes beyond what we can see. Central concepts in thermal science, such as energy, heat and temperature are not visible to the naked eye. These concepts permeate physics, chemistry, and biology as scientific areas of study, but also teaching of the corresponding subjects at multiple levels of education. As witnessed in the educational research literature through decades, learning thermal concepts continues to be a significant obstacle in science teaching across the world.

The learning and teaching of thermal science has traditionally relied upon the use of laboratory equipment, such as thermometers, to measure physical quantities, and to model such quantities using mathematical formalism. While these methods allow for the quantification of thermodynamic processes, they provide limited qualitative understanding of the phenomena.

The use of different tools to visualise such otherwise invisible phenomena is a potential approach to provide learners with a deeper conceptual understanding. For example, static visual representations such as graphs and diagrams can be used to communicate how different quantities relate to one another in different physical processes. With the emergence of computer-based tools, there are new opportunities to visualise what is too small, too large, or too complex in a dynamic way. For example, a simulation of molecular dynamics that allows for the adjustment of

J. Haglund (✉)
Department of Engineering and Physics, Karlstad University, Karlstad, Sweden
e-mail: jesper.haglund@kau.se

F. Jeppsson
Department of Behavioural Sciences and Learning, Linköping University, Norrköping, Sweden
e-mail: fredrik.jeppsson@liu.se

K. J. Schönborn
Department of Science and Technology, Linköping University, Norrköping, Sweden
e-mail: konrad.schonborn@liu.se

J. Haglund et al. (eds.), *Thermal Cameras in Science Education*, Innovations in Science Education and Technology 26, https://doi.org/10.1007/978-3-030-85288-7_1

parameters, such as temperature and pressure and different kinds of particles, allows for insight to be gained into thermodynamic processes at the microscopic level.

The introduction of Thermal Cameras in Science Education – the theme of this volume – presents the possibility to provide visual information of thermal processes in real time. Hand-held infrared (IR) cameras of different types have been found particularly attractive in science teaching. With graphical interfaces, and colour schemes that connect to the everyday perceptions of associating blue to cold and red to warm, the technology is user friendly and intuitive for learners of all ages. The immediate impression of seeing thermal phenomena visualised along a colour scale is often a very engaging and aesthetic experience. Furthermore, the technology is very versatile, which is expressed in the wide variety of applications in physics, chemistry, and biology teaching described in this volume. As prices continue to decrease, IR cameras are becoming a feasible option in science education, from preschool through university levels.

The potential use of Thermal Cameras in Science Education has received increasing attention during the last two decades. We are very proud that many of the educational researchers who have provided examples of how IR cameras can be used in science teaching and conducted empirical studies of students' use of the technology in learning science have contributed to this volume. In particular, we would like to acknowledge the early contributions to this field by scholars such as Michael Vollmer and Klaus-Peter Möllmann in Germany, and Charles Xie in the USA.

This volume is structured as three parts that communicate aspects of the fundamentals, educational research, and related use of Thermal Cameras in Science Education.

Part I provides an introduction to the physics and technology of thermal cameras. Vollmer communicates the fundamentals of thermal imaging that includes a description of thermal radiation as a physical phenomenon, and the principle of the technology of how thermal cameras of different kinds can be used to detect and visualise thermal radiation. In continuation of this theme, Oss compares IR camera data with theoretical models of heat transport.

The chapters comprising Part II present educational research on the use of thermal imaging in teaching. These contributions span across different science subjects and topics, as well as multiple levels of education. Samuelsson provides a synthesis of the literature on science education research with IR cameras. He describes how three types of papers, namely instructive, empirical and theory-informed studies have moulded this area of inquiry. Nordine, Weßnigk and Greinert show how thermal cameras can be used in pedagogical approaches for supporting a deeper understanding of energy in secondary school. In a pre-/post-test set-up they found that students' learning of energy conservation and dissipation was significantly higher when students used thermal cameras compared to teaching without thermal cameras. Loukomies, Makkonen, Juuti, and Lavonen presents a study on primary school pupils' emotions and experiences in relation to the use of thermal cameras in activities during a science lesson. The results emphasise the importance of recognising affective dimensions in young pupils' learning of primary science.

In continuing with a physics learning context, Åhman and Jeppsson describe a social semiotics analysis of pupils' practical work with IR cameras. The authors explore how thermal cameras can serve as a semiotic resource in supporting the understanding of heat and temperature. The closing chapter of Part II by Haglund, Folestam and Westlund moves away from physics teaching in exploring how upper-secondary students use thermal cameras in investigating animal temperature regulation. The work identifies opportunities for how IR cameras in combination with open-ended inquiry tasks can support learning of various concepts in animal physiology.

Finally, Part III gives examples of practical experiences of using thermal cameras in education, with a focus on how science teachers can utilise the technology in their teaching. In the first chapter of Part III, Vollmer presents the opportunities that smartphone accessory IR cameras offer, as well as their shortcomings, with respect to qualitative visualization and quantitative measurement. Planinšič, Nered, and Etkina have developed and tested practical activities with infrared cameras with the ambition to teach students to think like physicists. Wong and Subramaniam provide experiences from a wide range of activities using thermal cameras in physics, chemistry, and biology education. Finally, in consideration of exploring heat in public science centres, Lundin Palmerius and Schönborn describe the design and implementation of an augmented reality system for real-time visualization of heat transfer. In the application, thermal images of surfaces are projected onto the objects themselves as part of various hands-on activities.

The book ends with the chapter Conclusions and future outlook, where we as editors of the volume reflect on the contributions and what we see as the next steps with regard to how thermal cameras can be used in science education and educational research. The section is organised as a response to an overarching question: What are the current opportunities and challenges in introducing Thermal Cameras in Science Education, and what does the future hold?

The editors would like to heartily thank the contributing authors, and all teachers and students that have participated in their studies. Furthermore we would like to thank the authors that have provided comments on other chapters in the book as part of the internal review process, and the external reviewers of the volume. In addition, we gratefully acknowledge Ann-Marie Pendrill, Maja Elmgren, and Marcus Berg for providing expert comments on some of the chapters during the internal review process. We also thank Charles Xie for composing the Foreword to this book. Last but not least, we appreciate the efforts of Claudia Acuna and the editorial team at Springer for their support in putting the book together.

Our ambition is for this volume to provide readers with helpful insight and inspiration for educational researchers and practitioners in how thermal camera technology can be used as a pedagogical tool for meeting the challenge of learning and teaching demanding concepts in thermal science.

Part I
Thermal Imaging Technology and Physics

Chapter 2
Fundamentals of Thermal Imaging

Michael Vollmer

2.1 Introduction

Every object in the universe emits electromagnetic radiation, i.e. electromagnetic waves! The spectrum of the radiation depends only on two parameters, the absolute temperature T, measured in K and the material property emissivity. Most object temperatures in technology and nature vary between around 250 K and 3000 K. In this temperature regime, the spectral peaks in the wavelength representation range between around 12 μm and 1 μm, i.e. in the so-called thermal infrared (IR) range. The techniques to observe the emitted radiation with imaging methods is therefore called infrared thermal imaging – sometimes practitioners use the less precise term thermography which more implies a qualitative recording rather than the actual quantitative measurement technique. This chapter will briefly summarize the fundamentals of infrared thermal imaging. Much more extensive descriptions can be found in a recent textbook (Vollmer & Möllmann, 2018) and a large number of monographs or review articles (Budzier & Gerlach, 2011; Dereniak & Boreman, 1996; De Witt & Nutter, 1988; Holst, 1993, 2000; Moore, 2001; Rogalski, 2012; Wolfe & Zissis, 1993).

2.2 Electromagnetic Radiation

Electromagnetic (EM) radiation is described as electromagnetic waves. In the far fields (the only case relevant here) electric and magnetic fields are perpendicular to

M. Vollmer (✉)
Department of Engineering, University of Applied Sciences Brandenburg, Brandenburg an der Havel, Germany
e-mail: michael.vollmer@th-brandenburg.de

J. Haglund et al. (eds.), *Thermal Cameras in Science Education*, Innovations in Science Education and Technology 26, https://doi.org/10.1007/978-3-030-85288-7_2

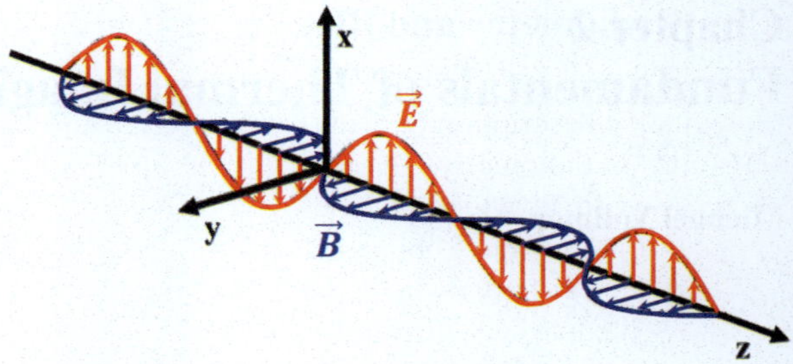

Fig. 2.1 Electromagnetic radiation resembles electromagnetic waves. They are characterized by the transient periodicity, defined by the frequency ν and the spatial periodicity, defined by the wavelength λ

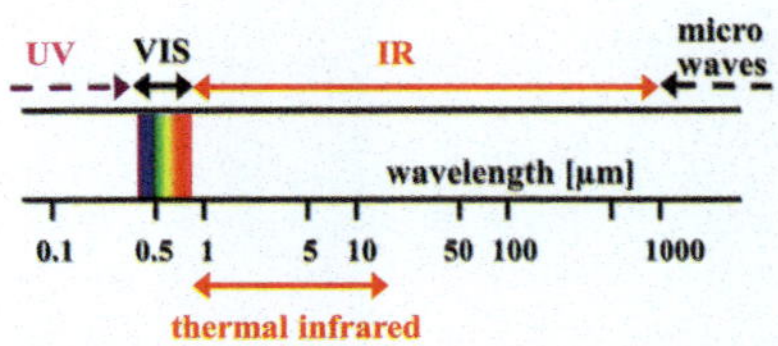

Fig. 2.2 Spectrum of EM radiation in the vicinity of the visible and infrared regions. IR imaging utilizes mostly the thermal infrared range with wavelengths between around 1 μm and 15 μm

the propagation direction. The simplest description starts with harmonically varying fields as shown in Fig. 2.1 which depicts a snapshot at fixed time. The spatial periodicity is the wavelength λ. A similar plot for the fields would result for fixed position as a function of time. The respective periodicity for one oscillation is the period, however it is more usual to use its inverse, the frequency ν. Wavelength is determined from frequency and the speed of propagation c as $\frac{c}{\nu}$. This is often more conveniently written as

$$c = \nu\lambda \tag{2.1}$$

For visible radiation c is called speed of light. In matter, $\lambda = \lambda_0/n(\lambda)$ and $c = c_0/n$ where the index 0 refers to vacuum and $n(\lambda)$ is the refractive index (see e.g. Bass, 1995; Hecht, 1998).

Frequency or wavelength are used to define special types of EM radiation. In nature and technology, waves spanning more than 15 orders of magnitude are relevant. Many of them can also be used for visualization purposes. Figure 2.2 depicts the types of EM radiation adjacent to the visible spectral range. Infrared is defined as having wavelengths between the red end of visible (VIS) radiation at $\lambda \approx 0.8$ μm, up to $\lambda = 1$ mm. The so called thermal IR, utilized in IR thermal imaging covers only the narrower range from around 1 μm to 15 μm (see below).

2.3 Radiometry and Blackbody Radiation

Quantitative description of electromagnetic radiation is possible using well defined quantities in radiometry. The most common one is the radiant power Φ also called energy flux in W which gives the total hemi-spherically emitted energy from the surface of an (opaque) object per time interval. The next important quantity is the excitance, i.e. the radiant power per emitting surface area $M = d\Phi/dA$ in W/m^2. Very similarly, the irradiance E is the energy flux incident on a given surface area. For all imaging applications, the most important quantity is the radiance L, which describes the emitted radiant power from an extended source per solid angle and projected source area in a given direction with respect to the surface normal of the object. In simple terms, radiance represents a measure of brightness of a source, observed in a given direction. If a source is perceived equally bright irrespective of observation direction, L is constant and the source is said to be a Lambertian radiator. Those sources are most important in daily life and technology. Consider a white screen or the tapestry in a room which is illuminated, e.g. by a projector. Due to the scattered light the screen/tapestry acts as light source. Observing the screen while moving around in the room, one is used to the fact that the perceived brightness stays the same. If this would not be the case – imagine that the brightness of tapestry would change when moving around – the result would be quite disturbing. Most objects which emit thermal radiation behave approximately as Lambertian radiators (more details on radiometry, see Vollmer & Möllmann, 2018).

As stated in the introduction, all objects in the universe emit radiation which depends on temperature and emissivity. In the simplest case the radiation only depends on temperature. The respective emitting objects are called black bodies. They are defined as follows:

> A blackbody absorbs every incident radiation, regardless of wavelength and direction. For given temperature and wavelength, no surface can emit more energy. It resembles an ideal Lambertian radiator.

Black bodies are idealizations but there are objects which come close. In science the best realizations are cavities with small holes (similar cavities are also known in everyday life. Think of a distant building with large rooms and only a small window is open and all internal lights are off. The window section looks very black. In the class-room one may use a shoebox covered by a lid with only a single small hole in one of the sides. For studies of blackbody radiation, the walls of the cavity are held at constant temperature. Then one may measure the emitted radiation from the hole as a function of wavelength while varying the temperature as only adjustable parameter. Precise studies were made at the end of the nineteenth century and in 1900 Max Planck was the first to correctly describe the spectrum of the emitted radiation, thereby – without intending to do so – introducing the era of quantum physics.

In modern language, the spectral excitance $M_\lambda(T)$, i.e. total radiant power into the hemisphere per area of a blackbody of given temperature T in wavelength interval $(\lambda, \lambda + d\lambda)$ can be written as

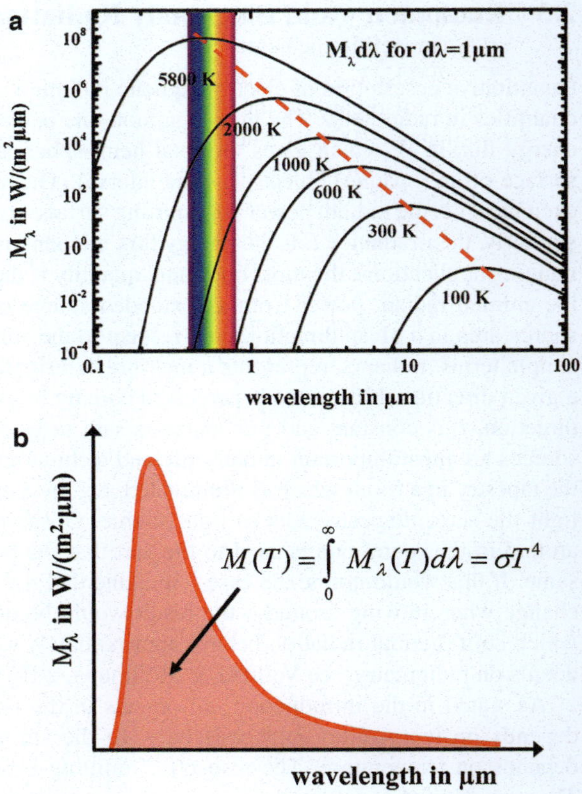

Fig. 2.3 (**a**) Spectral excitance of blackbody radiation for given temperatures as function of wavelength in double logarithmic plot. The broken line refers to Wien's law. (**b**) the integrated spectral excitance (shaded area) only depends on temperature and is expressed by the Stefan-Boltzmann law

$$M_{BB,\lambda}(T)d\lambda = \frac{2\pi hc^2}{\lambda^5}\,\frac{1}{e^{\frac{hc}{\lambda kT}}-1}\,d\lambda \tag{2.2}$$

The respective radiance would be $L_\lambda(T) = M_\lambda(T)/\pi$. Here, $h = 6.626 \times 10^{-34}$ J s is Planck's constant, $c = 2.998 \times 10^8$ m s^{-1} is the speed of light in vacuum, $k = 1.381 \times 10^{-23}$ J/K is the Boltzmann constant, λ is the wavelength of the radiation, and T is the absolute temperature of the blackbody given in Kelvin. Figure 2.3a depicts such blackbody spectra in a double logarithmic plot for temperatures between 100 K and around 6000 K. For convenience the VIS range is indicated by a colored band. Excitance $M(\lambda)$ increases with temperature at all wavelengths. The wavelength λ_{max} of maximum emission shifts to smaller wavelength with increasing temperature, indicated by the broken line. This shift is described by Wien's law

$$\lambda_{max} \cdot T = 2897.8\ \mu m\ K \tag{2.3}$$

For example, at $T = 5800$ K (temperature of the sun's surface layers), $\lambda_{max} \approx 0.5$ μm well in the center of the VIS range whereas at room temperature of around $T = 290$ K, $\lambda_{max} \approx 10$ μm.

Obviously, (disregarding the 100 K case) all spectra cover the IR range from 1 μm to well above 100 μm. Therefore, from radiometry alone, one could just guess that any IR camera should be sensitive between 1 μm and about 100 μm. There are, however, additional constraints for actually used wavebands which finally restrict IR imaging to the thermal IR range defined in Fig. 2.2.

The emission is highly nonlinear: the 6000 K source emits around 100,000 times more within a wavelength interval $d\lambda$ around the peak wavelength than the room temperature object. As a matter of fact, the total excitance, i.e. the spectral excitance integrated over the whole spectrum (Fig. 2.3b) is described by a simple law, the Stefan Boltzmann law

$$M(T) = \int_0^\infty M_\lambda(T)d\lambda = \int_0^\infty M_\nu(T)d\nu = \sigma T^4 \tag{2.4}$$

where $\sigma = 5.67 \times 10^{-8}$ W m^{-2} K^{-4} denotes the Stefan-Boltzmann constant. This means that if a sensor would detect the complete spectrum (which is never the case), its signal would depend on temperature to the fourth power.

2.4 Emissivity

Real objects in our world cannot be regarded as black bodies. Of course there are some realizations of near perfect black bodies at national metrology institutes, but mostly, one needs to deal with the corrections described by the object related parameter called emissivity. In brief, emissivity ε describes the ratio of the real emission of an object to its maximum possible emission if treated as black body. In other words, any real object emits radiation (here given as excitance) according to

$$M_\lambda(T)d\lambda = \varepsilon(\lambda)\ M_{BB,\lambda}(T)\ d\lambda \tag{2.5}$$

According to its definition, ε varies between near zero (no emission at all) and unity (black body). The best technical cavity blackbodies have ε above 0.999, the lowest values do have freshly created polished metals with ε around 0.01, so the span is quite large. Typical commercial blackbodies have $\varepsilon = 0.98$.

Figure 2.4a depicts three arbitrarily chosen examples of emissivity, first a black body ($\varepsilon = 1.0$), second a so called grey body with wavelength independent emissivity, here $\varepsilon = 0.6$, and third a selective emitter with strong variations as function of wavelength. Figure 2.4b shows the respective spectral radiance for T = 373K.

For any quantitative analysis of object temperature, one must know ε. Imagine a certain amount of radiation from an object of temperature T_1 is detected by a camera.

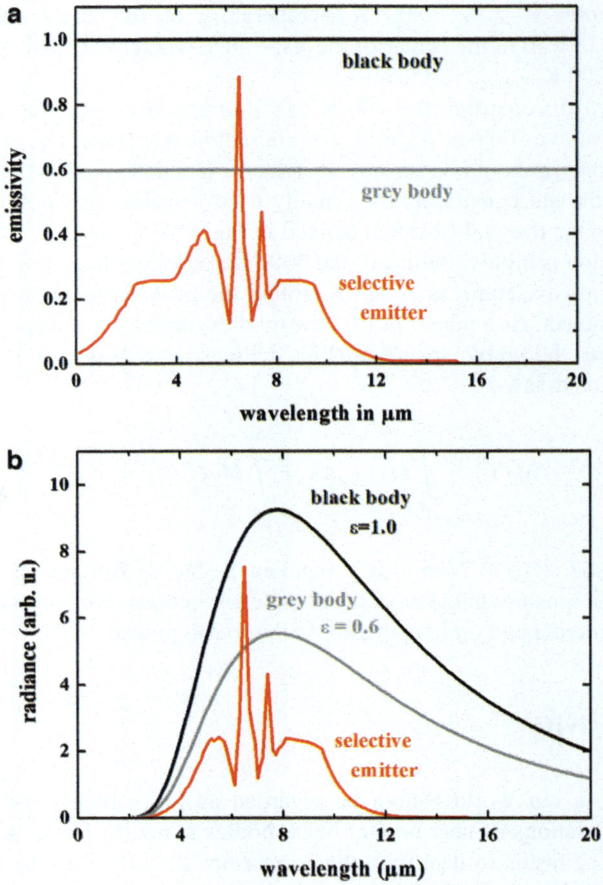

Fig. 2.4 The material property emissivity determines the ratio of emitted thermal radiation with regard to the blackbody radiation of an object of same temperature. If ε does not depend on wavelength one refers to grey bodies in contrast to wavelength dependent selective emitters

If the object is a blackbody, one may in principle (see radiometric chain below) derive this temperature T_1 using Eq. 2.2 integrated over the detection waveband and incorporating all necessary corrections. If however, the object of same temperature T_1 would be a grey object with an emissivity $\varepsilon = 0.1$ (i.e. 10%), the camera would detect only 10% of the previous amount of radiation. Assuming the previous blackbody object setting, this would result in a much lower temperature reading.

In contrast, we now assume that the exactly same amount of radiation as from the blackbody with T_1 is detected, this time however being due to an object with $\varepsilon = 0.1$ (i.e. 10%). The blackbody radiation of an object of same temperature T_2 would be a factor of 10 larger. This automatically means that the respective temperature T_2 will be much larger than T_1.

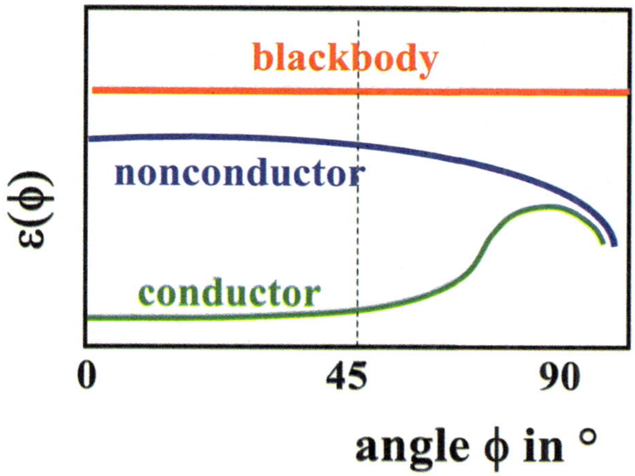

Fig. 2.5 Emissivity may depend on observation angle in case objects deviate from Lambertian radiators

For grey objects, the correction can be easily applied (substituting Eq. 2.2 by Eq. 2.5), for selective emitters, the analysis is more complex. In this case, quantitative analysis is still possible, using e.g. narrow band filters, broadband detection, however, is mostly done qualitatively. This holds in particular for gases and plastics. In nearly all practical teaching cases, it is sufficient to only refer to grey objects.

Unfortunately, even when restricting to grey bodies for a certain waveband and camera, emissivity is not always constant. Besides the dominant factor material (type of metal, insulator) it may depend on surface structure (rough, polished or weathered surface of same material), regular geometries (grooves, giving rise to interference effects), chosen waveband of a camera (SW, MW, LW), temperature (phase changes) and – very important – observation angle with respect to surface normal. Figure 2.5 depicts an example of the latter: emissivity of metals or dielectrics are nearly constant up to angles of around 45° but can appreciably change for larger angles. In this case, the objects do no longer behave like Lambertian radiators, which complicates the analysis. In practice, this means that one must avoid situations of large viewing angles, as can easily happen, e.g., in building inspections. The differences between conductors and nonconductors can be understood from polarization dependent reflectivities according to the Fresnel equations (see Vollmer & Möllmann, 2018).

2.5 Transmission of IR Radiation Through Air and Definition of Thermal IR Wavebands

When observing an object with an IR camera, there is always matter between the object and the camera detector (Fig. 2.6). Thereby radiation in the forward direction

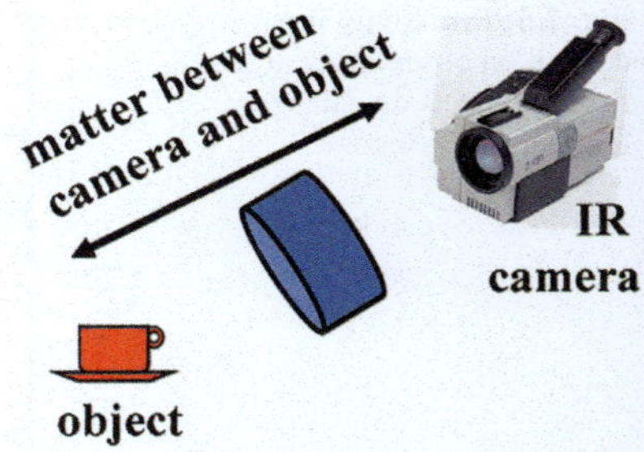

Fig. 2.6 Whenever a camera observes an object, emitting IR radiation, part of the radiation is scattered and /or absorbed, i.e. attenuated. In the simplest case, the radiation propagates through air

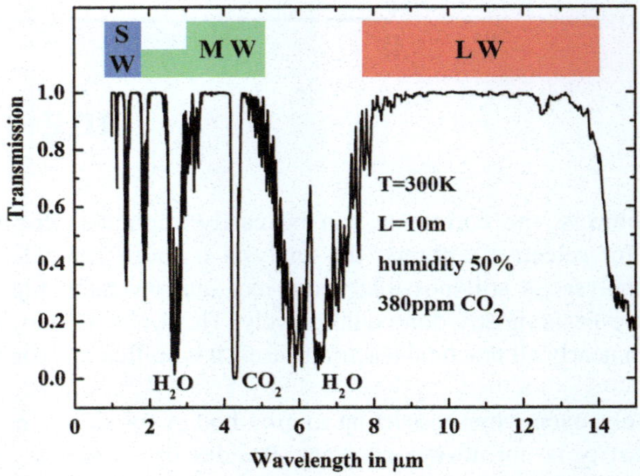

Fig. 2.7 Transmission spectrum through 10 m air at sea level. The atmospheric windows define the spectral bands used in IR imaging. They are indicated by colored bars

is attenuated due to absorption as well as scattering processes and the detector receives less radiation. Figure 2.7 shows an exemplary transmission spectrum through a 10 m horizontal path of air at sea level and ambient temperature of 300 K, 50% relative humidity and a CO_2 content of 380 ppm (in this overview spectrum, the result at the current CO_2 value of around 415 ppm does not differ appreciably).

The major constituents of air, N_2 and O_2 molecules, do not absorb in the thermal IR range, however, those molecules with permanent electric dipole moments, here in particular H_2O, and CO_2 may do so via excitation of rotations and vibrations, commonly denoted as ro-vibrational excitations. H_2O has absorption features nearly everywhere, but very strong absorption bands between 2.5 μm and 3 μm as well as between 5 μm and 8 μm and beyond 14 μm. Similarly CO_2 absorbs strongly close to 4.2 μm (the famous greenhouse gas absorption feature) and also beyond 14 μm. A very similar plot with an additional ozone feature is available for a radiation path from sea level to space, i.e. through the vertical atmosphere (Vollmer & Möllmann, 2018). As a result air has so-called atmospheric windows with little absorption. These are used in astronomy for looking through the vertical atmosphere and for

IR imaging when looking horizontally or at oblique paths within the atmosphere. Clearly, Fig. 2.7 shows that the least absorption occurs between around 8 μm and 14 μm (the LW thermal IR band) and 3 μm to 5 μm, (the MW thermal IR band). Please note, that the MW band includes CO_2-absorption. The band absorption is, however, well known, i.e. it can be accurately accounted for. For water vapor, corrections are more complicated as water vapor concentrations vary much more than those of CO_2.

Usually no cameras for wavelengths around the absorption features are available (disregarding special gas detection cameras). Quite recently, the SW band around 0.9 μm to 1.7 μm has gained more attention, particularly because very sensitive detectors are available. However, use of SW thermal cameras is restricted to high temperatures (Fig. 2.3a) and quantitative analysis is complicated by the absorption bands, i.e. additional careful calibrations are needed.

2.6 Measurement Process: Radiometric Chain

The measurement process of an IR camera is described by the so-called radiometric chain, which is visualized for the simplest case of opaque objects in Fig. 2.8. The investigated object emits blackbody radiation, which depends on the object temperature, modified by the emissivity ε and being characterized by the transmission $\tau(\lambda)$ through the atmosphere (red). In addition, the nearby ambient temperature background also emits IR radiation, part of it towards the observed object. The reflected part ($R = 1-\varepsilon$) also reaches the camera after passing through the atmosphere (green). Finally, whenever there is absorption within the atmosphere, there must also be emission (in thermal equilibrium, this is described by Kirchhoff's law) which means that there is also an atmospheric contribution (blue). In case objects are not opaque, there would be a fourth contribution from transmitted background radiation through the object.

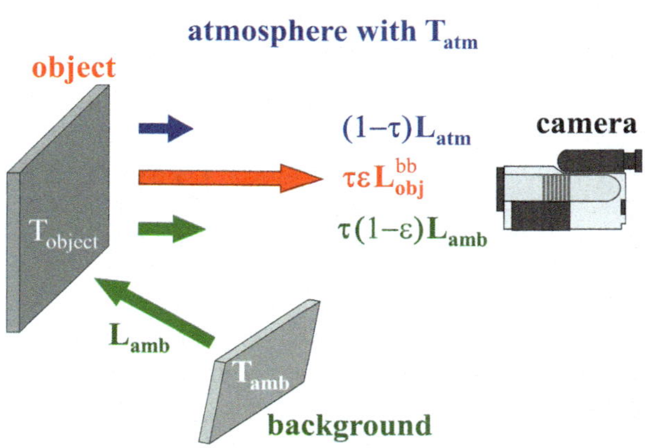

Fig. 2.8 Radiometric chain, describing the signals portions measured by an IR camera due to an opaque object

Obviously the information of object temperature is only contained in the object signal contribution. It is given by the black body object emission L_{obj}^{bb} integrated within the camera waveband (e.g. LW 8–14 μm or MW 3–5 μm) and modified e.g. by the transmission spectrum of the camera optics, etc. This signal is embedded in the other signals from the atmosphere and the background. In order to extract it correctly one needs to know ambient background temperature and object emissivity as well as transmission through the atmosphere. The latter involves knowing the distance to the object as well as relative humidity.

Once all necessary input parameters are known, a camera refers to previously performed and internally stored calibration curves. During the radiometric calibrations, the raw camera signals are measured for blackbody source objects as a function of temperature. Assuming correct user inputs, the camera then compares the measured raw signals of each pixel and attributes a temperature to them according to the internal calibration curve. The accessible temperature range can be extended by inserting attenuating filters in front of the detector.

2.7 Detectors and Optics

Once the waveband for detection is defined, it must be made accessible by first appropriate detector characteristics and second properties of the used optics in lenses and filters. Detection of infrared radiation is possible with various techniques, the most important being thermal detectors such as bolometers or detectors based on the photoelectric effect. A bolometer has a thin membrane which heats up while absorbing radiation. Thereby material properties change, in this case the electrical resistance, resulting in a measurable signal. Like all thermal detectors, heating up is governed by thermal time constants which depend on the mass of the material. Even for thin membranes, respective time constants are of the order of 10 ms, giving rise to camera frame rates of around 30 fps. In contrast, photoelectric detectors rely on excitation processes of electrons within semiconductor electronic bands. These are intrinsically much faster allowing much larger frame rates and high speed IR cameras. The choice of specific semiconductor materials enables optimized detectors for the LW and MW band of much higher sensitivity than bolometers. However, they need to be cooled and they are much more expensive.

Overall, most LW IR cameras operate with uncooled less expensive microbolometer focal plane arrays whereas the MW IR (e.g. InSb) and SW IR band (e.g. InGaAs) is almost exclusively dominated by photoelectric detector array cameras. The most expensive systems available in very high quality are based on HgCdTe detectors. Microbolometer detectors are usually broadband systems which means that filters (or the optics) restrict the waveband. Photoelectric detector cameras have an intrinsic long cut-on wavelength, but the short wavelength edge is often due to filters. The performance of detectors may be judged by a rough estimate for the radiant power changes of a single pixel upon small temperature changes. One finds about 2.5 nW/K, i.e. extremely small radiant power changes on

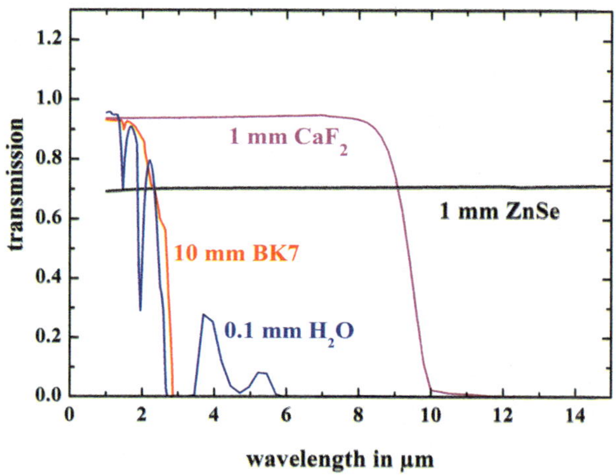

Fig. 2.9 Examples of optical transmission in the thermal IR range for four materials: 1 mm slab of CaF2, 1 mm slab of ZnSe, 10 mm slab of BK7 and 0.1 mm slab of pure water

the order of nW are still detectable. A very detailed discussion of detectors and their properties is given in Vollmer and Möllmann (2018).

Besides focal plane array detectors, cameras need suitable optics. For IR cameras, the materials must be transparent for the chosen spectral waveband region. Neglecting reflection losses (lens materials for IR camera are anyhow usually covered with anti-reflection coatings) the material transmission of solid slabs of thickness x is governed by the imaginary part of the refractive index n_2 according to

$$T(\lambda, x) = \frac{L(\lambda, x)}{L(\lambda, 0)} = e^{-\frac{4\pi n_2}{\lambda} x} \tag{2.6}$$

Data for refractive index of many materials are readily available (Palik, 1985). Figure 2.9 depicts external transmission spectra (i.e. those which still include reflection losses at the interfaces) from 1 µm to 15 µm for the common materials BK7 glass (red) and water (blue), as well as CaF_2 and ZnSe which are suitable for the MW and LW IR range. These data include reflection losses. Obviously, BK7 which is typical for many glasses, is only useful in the VIS and SW IR range. Liquid water has such strong absorption bands that 0.1 mm does not transmit any LW radiation and only a very small amount of MW IR. This immediately helps to answer the surprisingly popular question whether IR imaging allows to observe fish: the answer is no (and for the pedants: no if the fish is in water).

CaF_2 has high transmission up to around 8 µm wavelength whereas ZnSe can be used throughout the chosen wavelength region. Untreated, the latter has around 30% reflection losses. Therefore IR lens materials are usually all covered with anti-reflection coatings, optimized for the detection range. Those coatings work well for the design range, but show large variations of reflectivity as function of wavelength

Table 2.1 Typical specifications of commercial IR cameras

Property	Camera type: Low end	Average	High end
Example thermal resolution (pixels)	160 × 120	640 × 480	1280 × 1024
Detector	Bolometer	Bolometer, various photoelectric incl. QWIP	
Horizontal field of view (HFOV) due to optics	Variable, depends on optics		
Typical instantaneous field of view (IFOV)	Around 6 mrad	Around 1 mrad, better with telephoto lenses	
Frame rate	9–15 Hz	30 Hz	Several 10 KHz with subframing
Typical T range in °C	−20 to 400 °C	e.g. up to 2000 °C possible	
T accuracy	± 3 °C or ± 5%	± 2 °C or ± 2%	± 1 °C or ± 1%
Typical NETD @ 30 °C (Noise Equivalent Temperature Difference)	70 mK	30 mK	<20 mK
Emissivity settings	A few preset values	From 0 to 1, increments 0.01	
Other input parameters	None	Distance, T_{amb}, $T_{refl.}$, relative humidity	

for other spectral regions, including the VIS range. This explains the sometimes colorful appearance of IR lenses, which even depends on observation angle.

2.8 Some Specifications of IR Cameras

The behavior of the detector array and the optics determines the usual camera specifications (Table 2.1). For example, the spatial resolution is given by the number of pixels and the field of view, defined by the lens. The instantaneous field of view is the angular range imaged onto a single pixel. For example, a 55° FOV lens with 160 pixels yields an IFOV = 6 mrad. In contrast a telephoto lens of 20° FOV and 1000 pixels gives 0.35 mrad. Although this IFOV value is often given in camera specifications, the actual resolution is worse, as real lenses usually suffer from aberrations.

This also holds of course for IR lenses, in particular if they cover a large range such as 8 μm to 14 μm. As a result, IR lenses cannot provide ideal optical images where a given object area is perfectly imaged onto a single well defined pixel, only (see Fig. 2.10, left). Realistic IR imaging means that a given object area is imaged onto a larger than ideal area on the detector array (Fig. 2.10, middle). Conversely, a single pixel (pixel pitches as small as 12 μm have already been realized) does receive radiation not only from the ideally expected object area, but also adjacent object regions (Fig. 2.10, right). As a result, the spatial resolution of the camera is

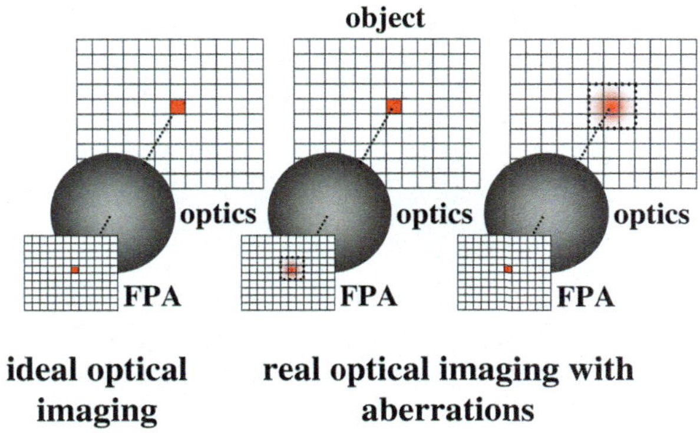

Fig. 2.10 Real optics aberrations reduce the spatial resolution

below the ideally expected done. As a rule of thumb, one needs at least an object size of 3 IFOV in order to get a reasonable temperature measurement for the central pixel (see also Vollmer & Möllmann, 2013).

The time resolution is determined by the detector response time. This is around 10 ms for microbolometer detectors and can be below 10 μs for photoelectric ones. In practice this means that once irradiation stops, the temperature signal decays exponentially and drops to (1/e) of the initial value after one time constant. Similar arguments hold for the onset of irradiation. It takes around five time constants before it drops below 1%. Usually one assumes that three time constants is a good compromise for effective signal rise or drop. This means that the detector integration time during a measurement should at least be three time constants. In the case of microbolometer cameras, this means around 30 ms. For photoelectric detectors its value is much lower, say around 30 μs. As the frame rate is 1/(integration time), we find a maximum frame rate for microbolometer cameras of around 30 Hz whereas photoelectric detector cameras can easily reach 30 kHz. Figure 2.11 visualizes the temperature errors, induced by observing moving objects with a low frame rate microbolometer camera. A 3 cm diameter rubber ball was heated in hot water to a temperature of around 340 K. When dropping it from 1 m height to the floor, it reached a velocity of about 2 m/s after falling 20 cm corresponding to a fall time of 0.2 s. The recording of an IR image is started when the ball reaches this position. During the integration time of the detector of 30 ms, the ball center drops around 6 cm, i.e. twice its diameter. Therefore it cannot be sharply imaged but shows a cometary tail. As a consequence, each pixel is only irradiated for part of the integration time and the measured temperature is below the actual one. Figure 2.11b shows that temperature differences of up to 22 K result even for such small velocities. Conclusion: a microbolometer camera will not give correct temperatures even for rather slow moving objects.

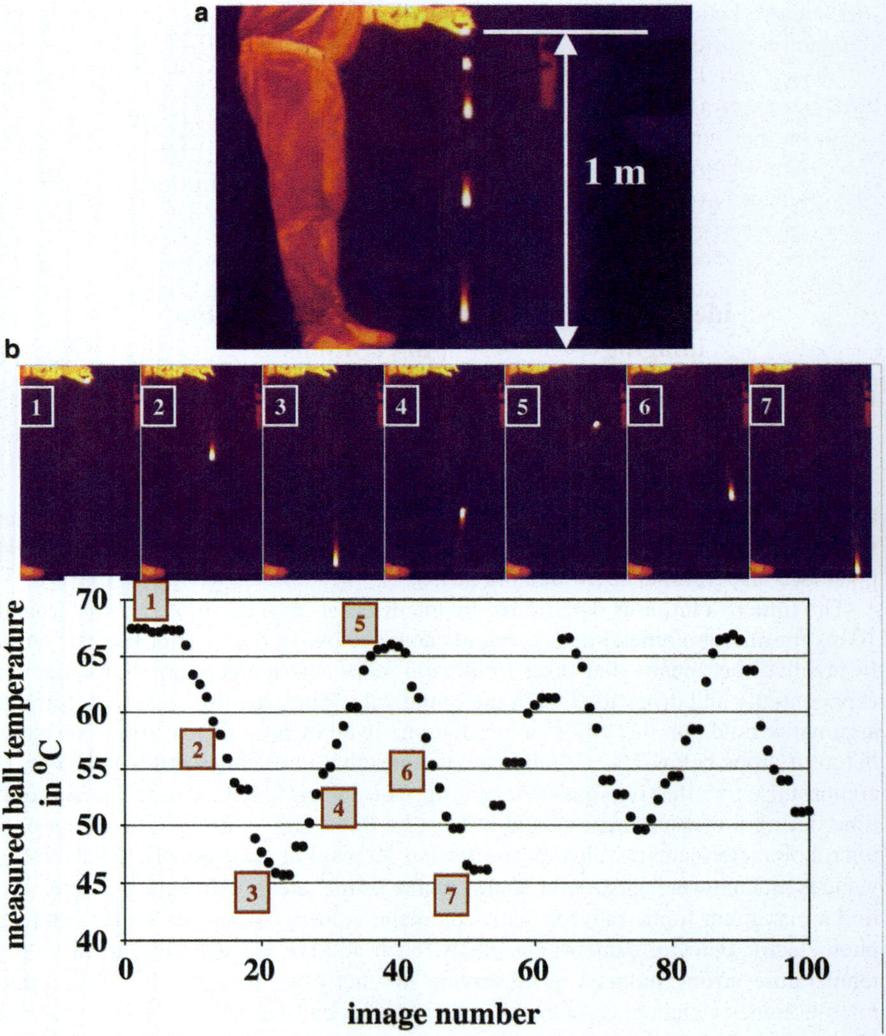

Fig. 2.11 The time resolution of an IR camera depends on response time of the detector and integration time of each camera pixel. Microbolometer detectors cannot accurately detect moving objects. (For details see text)

Finally, regarding temperature measurements, one must distinguish between the absolute and relative accuracy. For all thermometers, including IR cameras, absolute measurements have uncertainties of the order of several K. Table 2.1 shows, that the low end cameras have absolute uncertainties of ±3 °C or ±5% of the °C value, where the larger of the two values applies. The high end commercial cameras are at ±1 °C or ±1%. Only regularly recalibrated cameras at national metrology institutes can achieve better values. Fortunately, in physics, one is more often interested in temperature differences, only. This is described by the noise equivalent temperature difference (NETD), which is defined as the minimum difference between the temperature of a blackbody object and the temperature of a blackbody background at which the signal to noise ratio of the detector is unity. In this case the absolute temperature is irrelevant, only temperature differences are important. Typical low end products have rather large values of 70–100 mK, whereas high end cameras can well achieve below 20 mK, i.e. one may be able to distinguish minute temperature differences of below 0.02 K.

In addition to camera defined limits, there are additional user introduced uncertainties or errors due to insufficient or wrong input data. As discussed in the context of the radiometric chain one obviously needs to know emissivity, distance between object and camera, ambient temperature, atmospheric temperature as well as that of potentially reflecting nearby objects, and relative humidity. Low end camera software does not allow to account for most of them (see Chap. 9), the more expensive commercial products can account for all of them.

Whenever outdoor investigations are carried out, the list of potential problems and wrongdoings become excessively longer. These relate to time constants of investigated objects in relation to changes of the meteorological conditions such as wind, solar irradiation, shadows, clouds versus clear skies, precipitation and – very important – view factor effects of the surroundings. The latter accounts for the fact, that radiative transfer between objects depends on their relative orientation (more details, see Vollmer & Möllmann, 2018). Outdoor IR imaging needs a much deeper knowledge and often just leads to wrongly interpreted colorful images. Certified expert training courses are offered by many manufacturers and organizations which are highly recommended.

2.9 Temperature and Emissivity Contrast

After IR images are recorded, the problem is to properly interpret them. Mostly it is assumed that an IR image directly represents the surface temperature distribution of the object scene. This is incorrect! The detected radiation depends not only on temperature but also emissivity. Most object scenes contain many different objects of different materials and surface structures. As a result, contrast in nearly every IR image of such a scene does not only depend on temperature differences but also on the spatial emissivity distribution. As a rule, the observed IR image is therefore always a combination of two images, one representing actual temperature differences and the other emissivity differences. In short, the image is due to temperature contrast as well as emissivity contrast. Figure 2.12 shall illustrate the effect for an object scene with just four different geometric forms. The top image depicts a usual image of an IR camera with false color temperature scale. These images are always plotted for a single value of ε for all pixels. As shown, the detector receives four levels of radiation, least for the background, then successively more

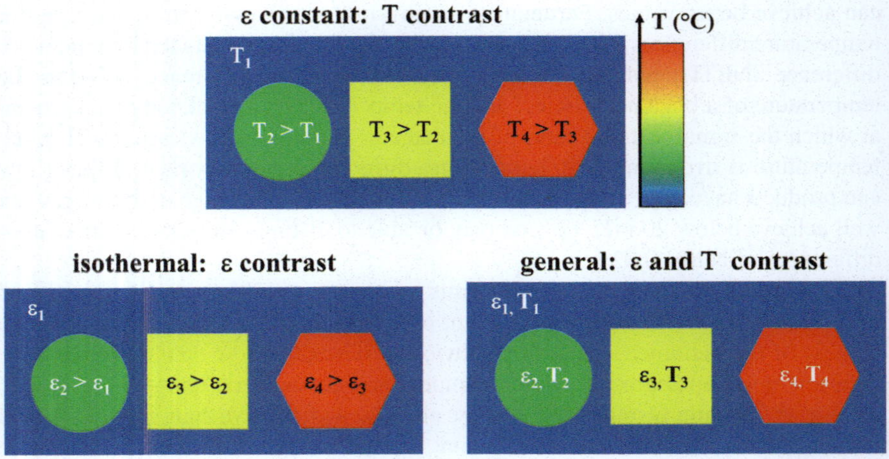

Fig. 2.12 Illustration of how emissivity and temperature affect the interpretation of IR images. (For details see text)

for circle, square and hexagon. As ε = constant, these differences are interpreted as being due to temperature changes. The internally stored calibration curve for the given input ε value results in the given temperature scale.

However, it could also well be (bottom left) that the object is isothermal and the different levels of radiation are just due to different emissivities. In this case emissivity increases from background via circle and square to the hexagon. Usually the truth lies in between the two extremes (bottom right): the various object parts have differences in temperature as well as emissivity. As they are shown in the same image, the allocated temperature scale is only valid for those parts of the image with the correct pre-set ε-value.

This example demonstrates that although T and ε contrast are easily separated theoretically, practical measurements are difficult. One experimental way to find ε contrast is to observe an object with emissivity contrast under isothermal conditions. One may place the object under study in an oven and achieve thermal equilibrium at $T_{\text{object}} = T_{\text{oven}} > T_{\text{ambient}}$. Removing the object from the oven and immediately recording its IR image will then show the emissivity contrast only, for the isothermal object. Similarly, hollow objects like Leslie cubes made from copper may be filled with hot water. The cube surfaces are isothermal due to the high heat conductivity of copper, therefore differences in the thermal images are only due to emissivity differences. Unfortunately, both methods are usually not feasible for real world objects which do not fit in ovens or cannot be filled with hot liquids.

The usual software tools do not allow to separately define an emissivity map for each pixel of the detector focal plane array (FPA) such that the resulting IR image would resemble a true temperature image. Rather one must focus on the interesting parts which are to be investigated and determine their emissivities and respective temperatures. Conclusion: quantitative analysis of IR images usually focusses on those parts of the image where emissivities are well known (for methods to determine emissivity experimentally, see Vollmer & Möllmann, 2018).

2.10 Pros and Cons of the Various Wavebands

Last but not least, an obvious question is which of the three spectral ranges, SW, MW or LW is best for IR imaging. Unfortunately the answer is not simple. First, there are radiometric arguments, then one may argue with properties of the various detectors which obviously also influence the cost of the cameras. Here, focus is on radiometric arguments. Figure 2.13a depicts part of Fig. 2.3a, i.e., spectral excitance

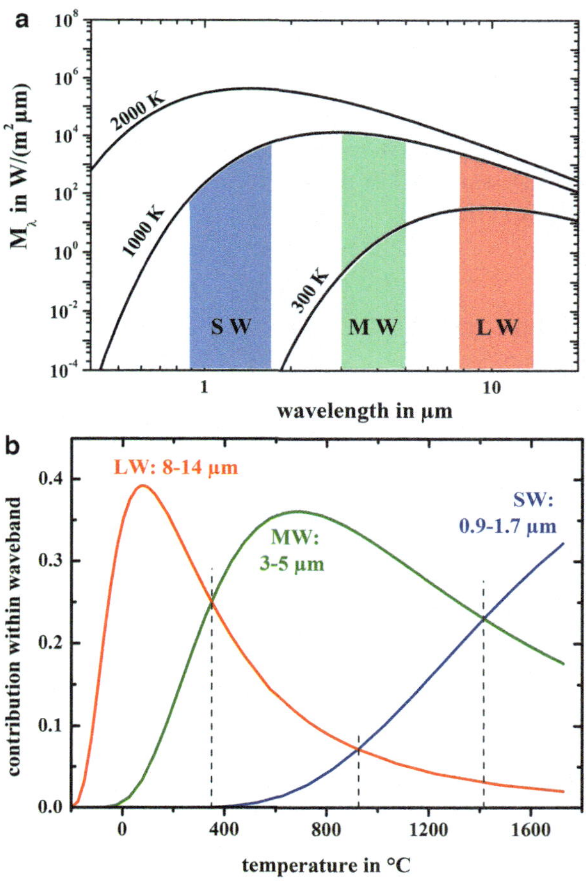

Fig. 2.13 (**a**) Total excitance for objects of three temperatures as a function of wavelength in double logarithmic plot. The exemplary colored area below the 1000 K curve resemble the detected radiation in the given wavebands. (**b**) the fraction of radiation in given wavebands as a function of temperature

for black bodies of 300 K, 1000 K and 2000 K again in a double logarithmic plot. Obviously the total amount of detectable radiation (Eq. 2.7) within the wavebands differ.

$$M_{band}(T) = \int_{\lambda_1}^{\lambda_2} M_\lambda(T)d\lambda \tag{2.7}$$

For 300 K, most of the radiation will be in the LW band, slightly less in the MW band and nearly nothing in the SW band. Figure 2.13b shows a linear scale plot of the portion of band radiation with respect to total radiation at a given temperature. Below 350 °C, the LW band receives the largest portion, from 350 °C to around 1420 °C, the MW band is the winner and above the SW band. This does not directly translate into band choice as, e.g., photoelectric detectors in the MW band regimes are more effective than microbolometer detectors in the LW band, compensating for the lower radiation at low temperatures. In addition, temperature contrast, i.e. radiation changes upon temperature changes rather favor shorter wavelengths for detection. The same applies to the effect of background reflections as well as emissivity uncertainties. In all of these cases, shorter wavelengths can produce better results.

Unfortunately, the best performance photoelectric detectors are also much more expensive than LW microbolometer focal plane arrays. Therefore the choice of a certain waveband is mostly decided by available resources rather than objective physics criteria.

References

Bass, M. (Ed.). (1995). *Handbook of optics* (Sponsored by the Optical Society of America). McGraw Hill.

Budzier, H., & Gerlach, G. (2011). *Thermal infrared sensors*. Wiley.

De Witt, D. P., & Nutter, G. D. (Eds.). (1988). *Theory and practice of radiation thermometry*. Wiley.

Dereniak, E. L., & Boreman, G. D. (1996). *Infrared detectors and systems*. Wiley.

Hecht, E. (1998). *Optics* (3rd ed.). Addison-Wesley.

Holst, G. C. (1993). *Testing and evaluation of infrared imaging systems*. JCD Publishing Company.

Holst, G. C. (2000). *Common sense approach to thermal imaging*. SPIE Optical Engineering Press.

Moore, P. O. (Ed.). (2001). *Nondestructive testing handbook, infrared and thermal testing* (Vol. 3, 3rd ed.). American Society for Nondestructive Testing, Inc.

Palik, E. D. (Ed.). (1985). *Handbook of optical constants of solids* (Vol. 1; Vol. 2, 1991; Vol. 3, 1998). Academic.

Rogalski, A. (2012). Review: Progress in focal plane array technologies. *Progress in Quantum Electronics, 36*(1–2), 342–473.

Vollmer, M., & Möllmann, K.-P. (2013). Characterization of IR cameras in student labs. *European Journal of Physics, 34*(6), S73–S90.

Vollmer, M., & Möllmann, K.-P. (2018). *Infrared thermal imaging – Fundamentals, research and applications* (2nd ed.). Wiley.

Wolfe, W. L., & Zissis, G. J. (Eds.). (1993). *The infrared handbook*. Infrared Information Analysis Center, Environmental Research Institute of Michigan.

Chapter 3
Thermal Infrared Imaging as a Bridge Between Mathematical Models and the Laboratory

Stefano Oss

3.1 Introduction

When, several years ago, I had my first occasion to "see the invisible", i.e. to paint in multicolor a temperature field thanks to a (very expensive, at that time) thermal infrared camera, I realized at once that my knowledge of the theoretical framework behind this world was amply inappropriate. Quite often, the syllabi of introductory courses to general physics at college level do not include, if not in a very soft manner, any reference to the study of thermal processes for energy transfer. The majority of students are more or less capable and ready to address the simplified arguments leading to the uniform, linear gradient of temperature across a planar ideal structure, so they can predict and understand, at least approximately, the money expense for winter heating or summer cooling of their apartments. Students could also be instructed on the distinction among the well-known three transport mechanisms relating to a temperature variation, i.e. conductive, convective and electromagnetic radiative processes. But no more than this. Yet, an even slightly deeper insight into the well-hidden secrets of this – indeed complicated – world could be very helpful and motivating. It is a personal convincement that entering this realm can be very effectively done (also) by adopting a thermal infrared camera as a discovering instrument and practical measurement device. Some reasons for this will be detailed in the following. Here, we anticipate that, besides technical issues from both experimental and theoretical viewpoints, a sensible approach to some of the abovementioned "non-standard" aspects of thermal processes, can be very rewarding in undergraduate studies just because of this technology which, as elsewhere presented and amply discussed (Vollmer & Möllmann, 2018), is

---

S. Oss (✉)
Department of Physics, University of Trento, Povo, Italy
e-mail: stefano.oss@unitn.it

J. Haglund et al. (eds.), *Thermal Cameras in Science Education*, Innovations in Science Education and Technology 26, https://doi.org/10.1007/978-3-030-85288-7_3

becoming more and more widespread and adopted for its relatively low cost and ease of use. Here we will not pursue the very important aspects about a proper usage of a thermal camera, since the operational strategies in this respect are addressed elsewhere in this book. Nonetheless, a word of caution should always be spent in order not to consider the infrared signal acquired by the micro bolometer of the camera as an automatic and faithful rendition of the actual temperature field. The steps between electromagnetic radiation entering the sensor and thermal properties of the body under exam are many and quite several of them could be tricky and critical to be properly treated.

We concentrate now on a series of mathematical sceneries, which stay behind a creative, sometimes unconventional practice of the thermal camera. The educational idea at the basis of this part of the book is that, despite sometimes quite demanding steps in the formal development of models pertaining thermal processes, even a pre-college teacher should be at least informed and aware of these mathematical aspects. This would allow an efficacious educational path and learning progress of the students, besides a rewarding attitude of the teacher. The subjects of this chapter are devoted to (*i*) the analogies between electromagnetic and thermal leading laws; (*ii*) the treatment of certain time-dependent thermal conduction processes; (*iii*) the combined convective-conductive contribution to heat transfer and (*iv*) the comparison between lumped and non-lumped systems in a thermal framework. All these subjects, after a proper and brief addressing in a mathematical perspective, will be considered also in an experimental thermal imaging context.

3.2 When Thermal and Electromagnetic Laws Face and Enforce Each Other

Transport processes belong to quite a large family which include, among others, electric current and heat. So, it could very well happen that certain similarities would manifest when comparing these two worlds (Kirchhoff, 1845; Kohlrausch, 1884; Higgins, 1956). This is indeed the case since diffusion equations can be written for both fields of temperature T and electric potential V as in the following expressions:

$$\nabla^2 T + \frac{\upsilon}{k} = \frac{c_H}{k}\frac{\partial T}{\partial t} \tag{3.1}$$

$$\nabla^2 V + \frac{\iota}{\sigma} = \frac{c_E}{\sigma}\frac{\partial V}{\partial t} \tag{3.2}$$

in which the heat capacity per unit volume c_H and the specific electric capacitance c_E are introduced along with the heat flux υ and electric current per unit volume ι.

Thermal and electrical conductivities are k and σ, respectively. These laws can be quite easily justified in terms of an energy balance over a control volume. Since this demonstration has solid grounds also in an educational perspective (Robertson & Gross, 1958; Oss & Zendri, 2015), it will be rapidly recalled here.

First, we focus our attention on a thermal problem in which the control volume v exchanges the energy δQ in order to modify its temperature by the amount dT. If the specific heat the material of the control volume is given by c_H, one has

$$\delta Q = c_H \; v \, dT. \tag{3.3}$$

In perfect similarity with the above expression one can also write

$$\delta q = c_E \; v \, dV \tag{3.4}$$

for the electric charge variation δq associated with the electric potential variation dV of a control volume with specific capacitance c_E. It is then possible to account for flow processes through a control surface A in terms of both thermal/electrical conductivities k/σ according to the usual expressions projected along the direction x perpendicular to the control surface:

$$\frac{\partial Q}{\partial t} = -kA \, \frac{\partial T}{\partial x} \tag{3.5}$$

$$\frac{\partial q}{\partial t} = -\sigma A \, \frac{\partial V}{\partial x}. \tag{3.6}$$

Combination of Eqs. (3.3–3.4) with Eqs. (3.5–3.6) allows one to write, after extending the notation to the three-dimensional space, i.e. after inclusion of the Laplace operator, the Fourier Eqs. (3.1–3.2) in absence of sinks and/or sources of electric/thermal energy, i.e.

$$\nabla^2 T = \frac{c_H}{k} \frac{\partial T}{\partial t} \tag{3.7}$$

$$\nabla^2 V = \frac{c_E}{\sigma} \frac{\partial V}{\partial t}. \tag{3.8}$$

This heuristic derivation is not meant to imply that the majority of time-dependent phenomena in electromagnetism can be described in this way. In particular, the Maxwell equations give partial differential equations for electromagnetic waves that are in the hyperbolic class of partial differential equations, and none of the differential equations here belong to this class. If one considers static

Table 3.1 Corresponding quantities for thermal and electrical transport phenomena

Thermal	Electrical
T (temperature, kelvin)	V (potential, volt)
u (power density, W/m^3)	ι (current density, A/m^3)
c_H (specific heat, J/(K m^3))	c_E (specific capacitance, C/(V m^3))
k (conductivity, W/(m K))	σ (conductivity, A/(m·V))
P (power, W)	I (current, A)

processes, i.e. independent of time, the Laplace equations for thermal/electric scalar fields are recovered in the usual form

$$\nabla^2 T = 0, \tag{3.9}$$

$$\nabla^2 V = 0. \tag{3.10}$$

These similarities can be organized and expressed through a systematic correspondence among electric and thermal quantities according to Table 3.1.

Overall, this means that, if one can visualize thermal fields for a given configuration of a thermal source, the temperature must behave exactly as the electric potential when the same geometry is adopted for an electrode arrangement characterized by certain physical and geometrical parameters. In other words, this would allow for the (indirect) visualization of the potential scalar field (whose values are very difficult – if not impossible – to be directly acquired) or, equivalently, the temperature map corresponds to the electric potential map. This idea can be further and operatively exploited through the use of a thermal camera with quite obvious intentions.

We show in Fig. 3.1, as an example, the infrared picture of two disks maintained at different and fixed temperatures via heating/cooling operated by Peltier cells in contact with the disks. The temperature behavior can be put in a numerical one-to-one correspondence with the potential field of two insulated disks uniformly and differently charged. The comparison can be pushed to quantitative levels by explicit solution of the static equations in terms of appropriate boundary conditions. In principle, it is thus possible to address the solutions of possibly demanding differential equations recovering some of their relevant features on a direct, visual basis for quite arbitrary configurations of electrodes. More generally, also certain non-stationary aspects could be visualized with time-lapsed videos in which the temperature varies with time due to conductive mechanisms whose electric counterparts are to be found in electric currents or charge flux.

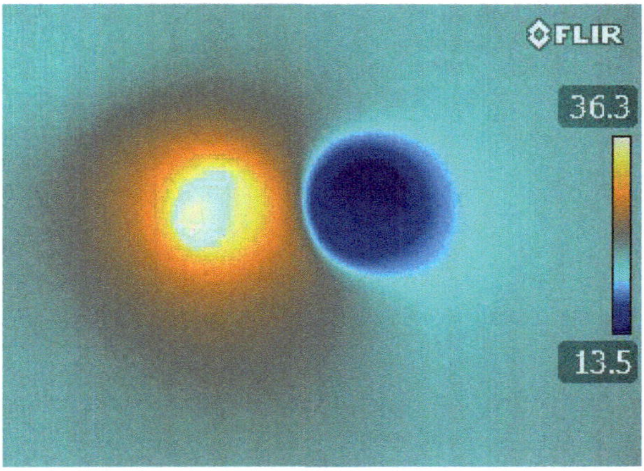

Fig. 3.1 Infrared image of two disks at different temperatures. The thermal field can be compared with the electric potential of two disks with different charge

3.3 When Time Varies: Thermal Movies

When speaking of non-stationarity in heat transfer processes, it is important to stress that the underlying mathematical apparatus can be a very challenging one. As recalled above, one has to deal with partial differential equations (PDE) of the Fourier type (which are usually classified as parabolic PDE). There exists a vast literature concerning these kinds of problems (Çengel, 2003; Incropera & DeWitt, 1990; Lienhard & Lienhard, 2019) and, as such, in this chapter no attempt will be made to reconstruct all these subjects. Here, we limit to address, as a specific example, the case in which thermal energy flows because of a non-equilibrium initial condition, i.e. one has a non-uniform temperature field in a given conducting object with well-defined geometrical features. More specifically, we sketch the typical case of a slab of uniform material whose sides are kept at given temperatures. The slab has an initial temperature which is different from the side temperatures. The PDE to be solved is

$$\frac{\partial T}{\partial t} = D\frac{\partial^2 T}{\partial x^2} \tag{3.11}$$

which is a rearrangement of Eq. (3.7): here we introduce the thermal diffusivity $D = \frac{k}{\rho c} = \frac{k}{C}$ for the material with thermal conductivity k, mass density ρ and specific heat c (or specific thermal capacity $C = \rho c$). The PDE is uniquely specified when initial conditions (I.C.) and boundary conditions (B.C.) are also provided.

As depicted in Fig. 3.2, the one-dimensional slab of length l has B.C. given by $T(x = 0, t) = T_L$, $T(x = l, t) = T_R$ (any time t) where L and R stand for the "left" and "right" side of the slab. For this same problem, the I.C. requires that $T(x, t = 0) = T_0$. The solution of Eq. (3.11) with these conditions is obtainable adopting either a Fourier analysis approach or the Laplace transform method (LTM). Both techniques are of course well-documented in the literature (Crank, 1975): here, we

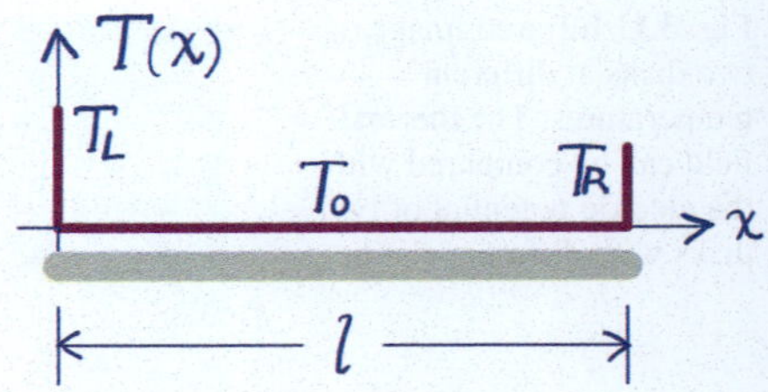

Fig. 3.2 Boundary/initial conditions coordinates of the uniform slab

limit for illustrative purposes to recall just some relevant steps of the LTM since it gives a somehow more physical insight to the settling of the solution.

The Laplace transform of $T\,(x,t)$ is defined as

$$L\left[T\,(x,t)\right] \equiv \tilde{T}(p) = \int_0^\infty e^{-pt}T\,(x,t)\,dt. \tag{3.12}$$

According to this definition, one recovers at once that

$$L\left[\frac{\partial^2 T}{\partial x^2}\right] = \frac{d^2\tilde{T}}{dx^2}, \quad L\left[\frac{\partial T}{\partial t}\right] = p\tilde{T} - T\,(x,0)\,. \tag{3.13}$$

On consequence, the PDE in Eq. (3.12), under the action of the Laplace transform becomes

$$\frac{d^2\tilde{T}}{dx^2} = \frac{1}{D}\left[p\tilde{T} - T\,(x,0)\right] = q^2\tilde{T} - \frac{T_0}{D}. \tag{3.14}$$

Here, we introduce the variable $q = \sqrt{p/D}$: the PDE has thus been transformed in the ordinary, second-order differential equation, Eq. (3.14), for the unknown function of x, $\tilde{T} = \tilde{T}_x(p)$, whose anti-transform is in turn the temperature field $T(x,t)$. The variable p has to be seen at this stage as a fixed parameter. Boundary conditions for this unknown function are $\tilde{T}_{x=0} = \frac{T_L}{p} = \frac{T_L}{q^2D}$, $\tilde{T}_{x=L} = \frac{T_R}{p} = \frac{T_R}{q^2D}$. It is straightforward to obtain the explicit, exact solution of Eq. (3.14) (notice that the I.C. of the original PDE is already included in the transformed differential equation). After some aesthetical retouch, one can write

$$\tilde{T}_x(p) = (T_R - T_0)\,\frac{\sinh qx}{p\sinh ql} + (T_L - T_0)\,\frac{\sinh q\,(l-x)}{p\sinh ql} + \frac{T_0}{p}. \tag{3.15}$$

This expression can also be rewritten in terms of infinite series starting from the expansions of the hyperbolic sine and of the binomial. The required expressions are

$$\frac{\sinh qx}{p\sinh ql} = \frac{1}{p}\sum_{k=0}^{\infty}\left\{e^{-q[l(2k+1)-x]} - e^{-q[l(2k+1)+x]}\right\},$$
$$\frac{\sinh q\,(l-x)}{p\sinh ql} = \frac{1}{p}\sum_{k=0}^{\infty}\left\{e^{-q[2kl+x]} - e^{-q[l(k+1)-x]}\right\}. \tag{3.16}$$

The reason of this rewriting is that, among the exactly known anti-transformed Laplace functions, one has

$$L^{-1}\left[\frac{e^{-qG}}{p}\right] = \text{erfc}\left(\frac{G}{2\sqrt{Dt}}\right) = 1 - \text{erf}\left(\frac{G}{2\sqrt{Dt}}\right) \tag{3.17}$$

where we introduce the (complementary) integral error or Gauss function $\text{erfc}(x) = 1\text{-erf}(x)$ which, as it is well-known, is obtained according to the definition

$$\text{erf}(x) = \frac{2}{\sqrt{\pi}}\int_0^x e^{-t^2}dt. \tag{3.18}$$

Finally, in terms of the error function and of the initial temperatures, the explicit expression of the temperature field can be written as the anti-transform of Eq. (3.15),

$$T\,(x,t) = T_0 + (T_R - T_0)\sum_{k=0}^{\infty}\left\{\text{erfc}\left[\frac{l(2k+1)-x}{2\sqrt{Dt}}\right] - \text{erfc}\left[\frac{l(2k+1)+x}{2\sqrt{Dt}}\right]\right\} +$$
$$+ (T_L - T_0)\sum_{k=0}^{\infty}\left\{\text{erfc}\left[\frac{2kl+x}{2\sqrt{Dt}}\right] - \text{erfc}\left[\frac{2l(k+1)-x}{2\sqrt{Dt}}\right]\right\}. \tag{3.19}$$

The temperature field for this problem, Eq. (3.19), can be computed numerically for assigned values of the relevant physical parameters as a function of time and space coordinates. A typical outcome is shown in Fig. 3.3: the initial behavior (step-function) of temperature changes gradually with time towards a stationary equilibrium, linear dependency along the slab.

Despite the intimidating aspect of the LTM result, several approximations can be done with the aim of simplifying, also in an educational framework, the comparison of the Fourier heat law with experimental (infrared) measurements. A first possibility is to look for relatively large arguments of the complementary error function erfc in Eq. (3.19), since this lead to negligible contribution to the terms of the infinite series. This is a better and better approximation for relatively shorter time intervals: more specifically, when $t < l^2/D$, only the first terms ($k = 0$) of the series survive for any value of the x coordinate. In this case, Eq. (3.19) simplifies to

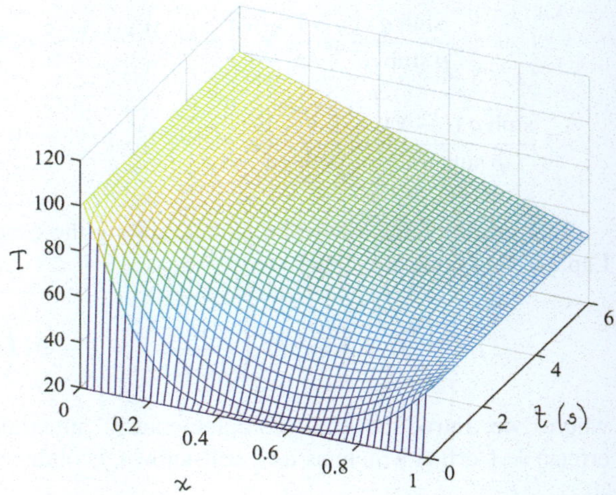

Fig. 3.3 Temperature field of the homogeneous slab with initial step-conditions

$$T(x,t) \cong T_0 + (T_R - T_0)\left\{\text{erfc}\left[\frac{l-x}{2\sqrt{Dt}}\right] - \text{erfc}\left[\frac{l+x}{2\sqrt{Dt}}\right]\right\}$$
$$+ (T_L - T_0)\left\{\text{erfc}\left[\frac{x}{2\sqrt{Dt}}\right] - \text{erfc}\left[\frac{2l-x}{2\sqrt{Dt}}\right]\right\}, \quad t < \frac{l^2}{D}. \tag{3.20}$$

This last expression can be further simplified when the temperature is acquired not too far away from the origin of coordinates along the slab, i.e. for $x < l$, in which case, always for not too long times, one obtains

$$T(x,t) \cong T_0 + (T_L - T_0)\,\text{erfc}\left[\frac{x}{2\sqrt{Dt}}\right], \quad t < \frac{l^2}{D}, \quad x < l, \tag{3.21}$$

which is an extremely simple dependence in x, t of the temperature field to be adopted in the study of some practical situations. We cite, as an example, the possibility of acquiring the temperature field of a solid rod with one of its ends maintained at a given "hot" temperature starting from an initial condition in which the rod is at the environment "cold" temperature. The opposite end of the slab can be held at another, different temperature or it can be considered as a non-conducting point. In both cases, if one considers small displacements from the "hot" side, for relatively short intervals of time Eq. (3.21) holds. So, one has a straightforward mathematical rendition of this easy experiment which can be realized in a very effective way through a thermal visualization of the thermal field, i.e. as a simultaneous dependence in x and t. We show in Fig. 3.4 temperature behaviors along a metal bar and the comparison with Eq. (3.21) in which the diffusivity D has been fitted to its best-approximating value against the observed data (Oss, 2020a).

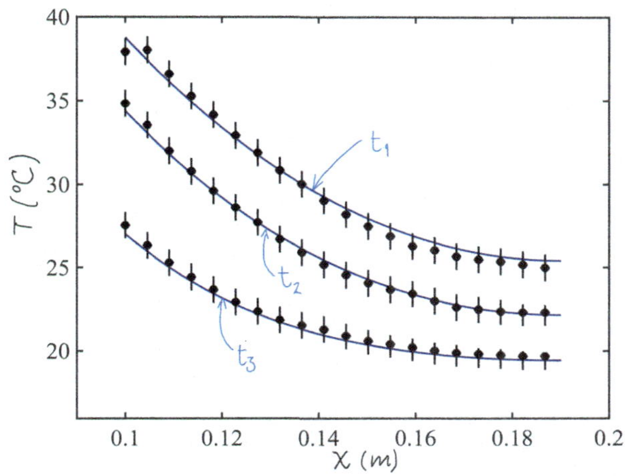

Fig. 3.4 Temperature behaviors along a metal rod with one end at fixed temperature. Values taken at 3 different instants of time ($t_1 > t_2 > t_3$)

3.4 When Air Matters

Heat transfer via convective processes is usually addressed at introductory levels by simply enunciating the Newton empirical law for the heat transfer rate

$$\dot{Q} = hS\,(T_B - T_\infty) \tag{3.22}$$

when a body with a surface S at a temperature T_B is exposed to an external (fluid) environment at temperatureT_∞. The parameter h is known as the heat transfer coefficient and it is measured in W/(m^{2}K). This model is only apparently a simple one: the parameter h is not constant but it changes in very intricate, often unpredictable ways as a function of many characteristics of the problem at issue. Fluid speed, effects of gravity, relative orientation of the heating/cooling body, the either natural or forced origin of the fluid motion are all relevant features to be accounted for when dealing with convective heat transfer. Despite these technical difficulties, convective processes are extremely important and, as such, they should be addressed in a somewhat rigorous fashion also in an introductive approach.

An interesting aspect is related to the combined role of conduction and convection when temperature gradients involve a solid body in contact with a fluid. The substantial mechanism driving these phenomena is of conductive origin (i.e. it should be interpreted as the energy exchange which occurs because of collisions at atomic/molecular scale). The mesoscopic mobility of fluid particles is then activated by the temperature/pressure/density non-uniform distributions. It is thus convenient, mainly in an educational perspective, but not only, to introduce the hybrid model in which a conductive mechanism is somehow affected by a simultaneously occurring convective heat transfer (Oss, 2020b).

Toward this aim, we consider thermal conduction through a solid plate whose ends are characterized by well-defined boundary conditions in a stationary (time-

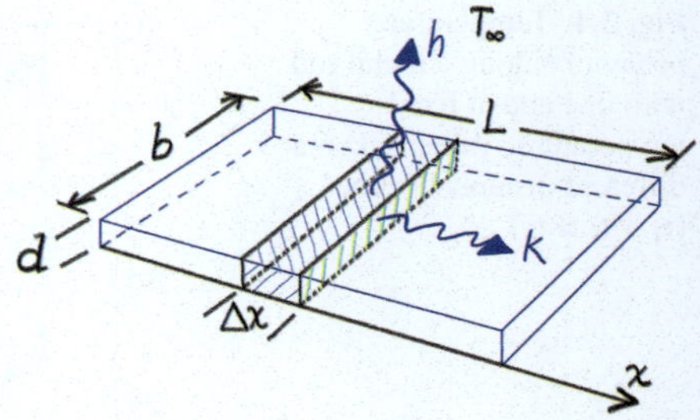

Fig. 3.5 Convective (h) and conductive (k) heat fluxes through a thin plate

independent) configuration. The plate, with length L, width b, thickness d and thermal conductivity k has a finite surface which is exposed to the external fluid environment (air, for instance) which can, because of its temperature gradient within the plate, exchange energy with it according to the simple Newton's law expressed by Eq. (3.22). We thus have a combined conduction-convection energy flux which, for a given control volume $\Delta V = \Delta x \cdot d \cdot b$ (see Fig. 3.5), can be expressed as

$$-k\,(d \cdot b)\,\frac{\Delta T}{\Delta x} + h\,(2b \cdot \Delta x)\,(T - T_\infty) = 0; \tag{3.23}$$

the factor 2 is required because there are two equal surfaces contributing to the convective flux. In this expression we put at the same footing the conductive (k) and convective (h) contributions to the energy balance. Eq. (3.23) can be reorganized in differential form as

$$\frac{d^2T}{dx^2} = \mu^2\,(T - T_\infty)\,, \quad \mu = \sqrt{\frac{2}{d}\frac{h}{k}}. \tag{3.24}$$

The ordinary differential Eq. (3.24) has as a straightforward general solution in the expression

$$T(x) = T_\infty + C_1 e^{+\mu x} + C_2 e^{-\mu x}. \tag{3.25}$$

Particular solutions are of course obtained after applying specific boundary conditions to satisfy the physical configuration at issue. Here, we consider two examples. In both cases, the left side of the plate is kept at a fixed "hot" temperature T_{H}. In the first case, also the right end of the plate is kept at a fixed temperature T_R. In the second case, there is no heat flux at the right end of the plate, as it happens for example when the surface of the terminal side is very small (d is small, yet b is not in order to allow for a relatively large longitudinal surface of the plate). The boundary conditions for these two configurations are thus given respectively by

$$T\,(x=0)=T_H,\quad T\,(x=L)=T_R; \tag{3.26}$$

and

$$T\,(x=0)=T_H,\quad \frac{dT}{dx}\,(x=L)=0. \tag{3.27}$$

The corresponding temperatures are readily obtained after substitution of Eqs. (3.26) and (3.27) in (3.25). The final results, after some rearrangement, are

$$T(x)=T_\infty+\frac{(T_R-T_\infty)\sinh\mu x+(T_H-T_\infty)\sinh\mu\,(L-x)}{\sinh\mu L} \tag{3.28}$$

and

$$T(x)=T_\infty+(T_H-T_\infty)\,\frac{\cosh\mu\,(L-x)}{\cosh\mu L}. \tag{3.29}$$

We recall that the coefficient μ expresses the relative contributions of the convective and conductive mechanisms. For very small, negligible convection ($\mu\to 0$) the expressions above reduce, within the first order in μ, respectively to

$$T(x)\cong T_H+\frac{x}{L}\,(T_R-T_H) \tag{3.30}$$

and

$$T(x)\cong T_H. \tag{3.31}$$

We notice that Eq. (3.30) represents the expected linear variation of temperature along a conductive medium whose ends are kept at fixed temperatures T_R and T_H, whilst Eq. (3.31) predicts that, in absence of convective processes, the equilibrium temperature of the conductive plate with one insulated end assumes the left side value. In presence of a non-negligible convection-to-conduction ratio, these curves depart from their linear behavior in x to follow Eqs. (3.28) and (3.30).

These kinds of behaviors can be observed directly through infrared imaging: an example of the fair agreement of this simple model with the experiment is shown in Fig. 3.6. The metal plate is heated at one end with an electric resistance and its surface (after having been painted with matt back paint to ensure a relatively high emissivity) is photographed with the thermal camera. Temperature profiles are obtained from the infrared picture with standard analysis software. Different values of the coefficient μ can be assigned by varying the convective strength h which, in turn, changes because of different speeds of the air which is blown on the exposed surface. It is also worth mentioning that this setup is used as a standard way to provide air cooling of electronic devices. So, the study of these processes has a relevant technological counterpart besides its educational appeal.

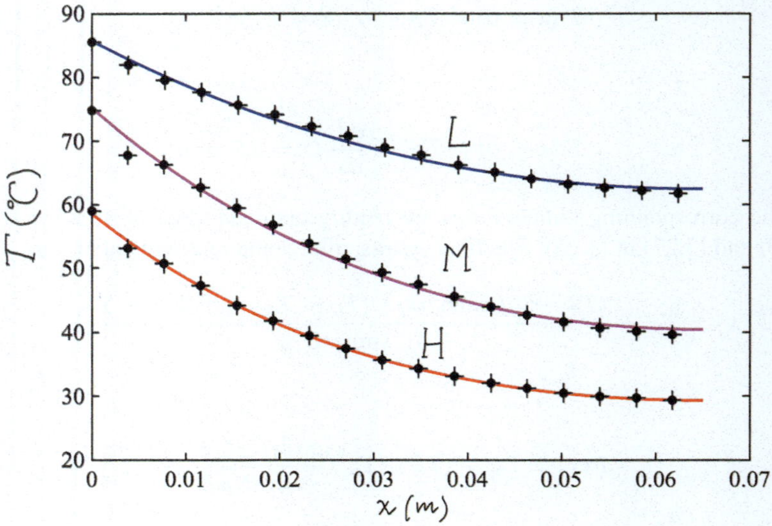

Fig. 3.6 Temperature values along a metal plate with high (H), medium (M) and low (L) speed air circulation. Infrared observed data and the theoretical model are compared

3.5 When the Thermal Camera Looks Inside a Conducting Body

As hinted to in the paragraph above, when a solid body with a given, uniform initial temperature is exposed to the thermal contact with an outside colder or warmer fluid environment, it undergoes a cooling/heating process, which can be very different in function of various parameters. Critical dependencies are those associated with the size and shape of the body, its thermal conductivity and the convection effectiveness of the external medium. In this paragraph, we briefly address the comparison between two case studies in which the solid body has a uniform/non uniform temperature distribution within itself. We then study the time dependency of the temperature field in these two cases. The distinction between uniform/non-uniform temperature fields inside a body subject to a thermal non-equilibrium condition is usually referred to as the lumped/non-lumped (or concentrated/non-concentrated) behavior. As quite typical examples of lumped bodies one considers small objects with a relatively high thermal conductivity embedded in a poorly convective environment. With these conditions, a thermal gradient from the outside propagates quickly and uniformly inside the body. On the contrary, if the internal conductivity is relatively small compared with the external convection, and/or the body is large, temperature variations are not uniformly distributed, i.e. they change not only as a function of time but also in different positions of the bulk of the body. Several textbooks, quite appropriately, suggest the example of a cake taken off an oven (or a chicken placed in the oven) with their uneven distributions of temperature to visualize a non-lumped behavior or, as an example of lumped system, yet staying in the kitchen, the cooling of an initially hot cup of coffee whose temperature is approximately uniform in its small volume during the cooling process. The case of the lumped (or concentrated) behavior is readily discussed within the framework of the simple Newton law, Eq. (3.22). One has just to establish the balance of ingoing and outgoing energy in the form

$$dQ = C\,dT = mc\,dT = hS\,(T - T_\infty)\,dt \tag{3.32}$$

which is readily integrated with the initial condition $T(t = 0) = T_0$ leading to

$$T(t) = T_\infty + (T_0 - T_\infty)\,e^{-\frac{hS}{mc}t}. \tag{3.33}$$

This last expression is the well-known exponential cooling (or heating) law in which one introduces the time-constant for the process $\tau = mc/(hS)$: a relatively fast heat transfer is expected for strong convection, large exposed area and/or small heat capacity. It is also customary to characterize the dimensions of the body through the parameter (characteristic length) $L = V/S$ (V is the body's volume). In this way, one can express the time constant as $\tau = \rho c \frac{L}{h}$ (ρ is the mass density). It is possible to introduce a quantitative/effective criterion for establishing whether the system under consideration is behaving as a lump of cooling/heating matter of if it is needed to consider a non-uniform temperature field inside the body. This criterion is based on the comparison of the thermal resistances for the internal (conductive) and the external (convective) processes at play: more explicitly, one introduces the so called Biot (dimensionless) number (Çengel, 2003)

$$Bi = \frac{L/k}{1/h}. \tag{3.34}$$

For small Biot numbers ($Bi < 0.1$), the lumped approximation holds (small body, large conductivity, poor convection) in the sense that temperature variations inside the body are smaller than 5%. For larger Biot numbers one has to address the full problem of the heat transfer as regulated by the partial differential Fourier equation, Eq. (3.11). Here, we consider as a specific example and application, the heating/cooling of a non-lumped body with cylindrical symmetry. It could be adopted as a sensible model of the cake extracted from the oven: yet, differently from the cake, in this model we assume that the heat flux is directly radial and no energy exchange occurs through the top/bottom bases of the cylinder which, in fact, is taken with infinite height. The Fourier equation has to be adapted to the cylindrical radial coordinate r according to

$$\frac{\partial T}{\partial t} = D\frac{1}{r}\frac{\partial}{\partial r}\left(r\frac{\partial T}{\partial r}\right). \tag{3.35}$$

The boundary conditions for this equation impose that the radial flux will be zero at the center of the cylinder whilst it will respond to the temperature gradient imposed by the convective process going from inside to outside:

$$\left(\frac{\partial T}{\partial r}\right)_{r=0} = 0, \quad \left(\frac{\partial T}{\partial r}\right)_{r=R} = \frac{Bi}{L}\left[T_\infty - T\left(r = R\right)\right]. \tag{3.36}$$

These conditions are assumed to hold for every instant of time. One has also as initial constraint, for any radial coordinate $r < R$, the condition

$$T\left(r, t = 0\right) = T_0 \tag{3.37}$$

which fixes the (initially uniform) temperature of the cake before the heat transfer begins. The partial differential Eq. (3.35) with its boundary and initial conditions, Eqs. (3.36) and (3.37), can be exactly solved in terms of transcendental mathematical functions: differently from the previous example, here we sketch the solution based on the Fourier analysis which refers to the separation of space and time coordinates to obtain ordinary differential equations parametrized in an eigenvector/eigenvalue framework (Crank, 1975). When applied to the cylindrical geometry, the Fourier method provides for the temperature field the following expression:

$$T\left(r, t\right) = T_\infty + \left(T_0 - T_\infty\right)\sum_{k=1}^{\infty} A_k e^{-w_k^2 \frac{D}{R^2}t} J_0\left(w_k \frac{r}{R}\right), \tag{3.38}$$

in which J_0 is the first-kind, zeroth order Bessel function and w_k are the roots of the characteristic equation

$$w_k J_1\left(w_k\right) / J_0\left(w_k\right) = Bi. \tag{3.39}$$

In this expression the first order Bessel function of the first kind, J_1, is introduced. The numbers A_k in Eq. (3.38) are finally obtained according to this equation:

$$A_k = \frac{2 J_1\left(w_k\right) / w_k}{\left[J_0\left(w_k\right)\right]^2 + \left[J_1\left(w_k\right)\right]^2}. \tag{3.40}$$

The infinite series in the temperature field (3.38) can be truncated to the first term, thus reducing the final result to a relatively simple expression, for sufficient long intervals of time since the beginning of the heat transfer process. More precisely, one introduces as a characteristic, dimensionless time, the so called Fourier number *Fo* for the infinite cylinder of radius *R* and diffusivity *D*:

$$Fo = D\frac{t}{R^2}. \tag{3.41}$$

A detailed computation (Çengel, 2003) leads to the result that the first term of the summation in Eq. (3.38) ensures an error less than 2% when $Fo > 0.2$ or, in terms of actual time, $t > 0.2\ R^2/D$. The associated, single-term law is explicitly given by

$$T\ (r, t) \cong T_\infty + A\ (T_0 - T_\infty)\ e^{-w_0^2 \frac{D}{R^2} t}\ J_0\left(w_0 \frac{r}{R}\right) \tag{3.42}$$

in which the constant w_0 depends on the Biot number for the problem at issue.

A relevant feature of the solution based on the Fourier technique is that time and radial dependencies are completely and explicitly factored out. This allows a simple tracking of the non-uniform distribution of temperature inside the body at any given instant of time. So, it would be very convenient to acquire experimental information concerning the internal temperature behavior in the non-lumped system. Thermal infrared imaging can be useful for this aim. As suggested in Oss (2021), it is possible to monitor the non-uniform cooling/heating inside of a cylindrical body caused by convective external mechanisms and sustained by conductive internal processes. One can take infrared images of the basis of a solid cylinder (other shapes are of course permitted but in that case the mathematical development of the theoretical model could become much more complicated) with its lateral surface in contact with cooler/warmer water (a liquid fluid ensures a non-negligible convective effect and a relatively uniform radial heat flux for not too large temperature gradients). Looking at the emerged basis of the basis is equivalent, to a good extent, to observe (and eventually acquire) thermal data inside the cylinder. This becomes more and more effective when the temperature of the cylinder remains not too dissimilar from that of the outside environment (i.e. the air in the room).

It is convenient to put the cylinder initially at room temperature in a bath of cold water: doing so, the dominant contribution to the heat flux will be that from/to the lateral (wet) surface of the cylinder, which is exactly what one needs to reproduce in order to realize the model herewith discussed. It is possible to change the material of the cylinder, i.e. to vary the conductivity and, as a consequence, the Biot number, Eq. (3.34), in which the convective parameter h is unchanged since the fluid is always the same; also the characteristic length L does not vary if we use cylinders with equal sizes made of different materials. It is then permitted a direct, visual inspection of the lumped (for good thermal conductors) and non-lumped (for good insulating materials) behaviors based on the analysis of pictures/movies taken with an infrared thermal camera.

We show in the upper panel of Fig. 3.7 a series of snapshots showing the radially uniform cooling of a conducting cylinder (which turns out to follow a simple exponential law of temperature decrease as a function of time, according to Eq. (3.42)). In the lower panel of Fig. 3.7 a similar sequence of frames shows, as a function of time, the temperature distributions at the basis, i.e. "inside" a poorly conducting cylinder (here made of Plexiglas). The non-uniformity of the radial

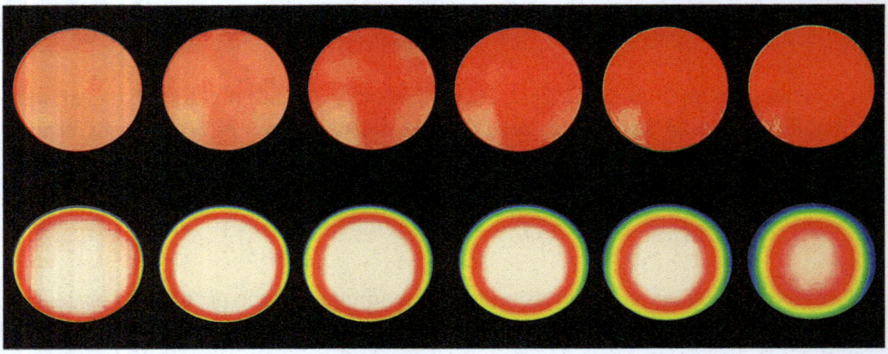

Fig. 3.7 (Top) Infrared snapshots at various instants of time of the basis of a conducting (brass) cylinder during its cooling in cold water as an example of lumped behavior. (Bottom) Infrared snapshots at various instants of time of the basis of an insulating (Plexiglas) cylinder during its cooling in cold water as an example of non-lumped behavior

distribution is evident and it can be associated with the r dependence of the zeroth-order Bessel function of the first kind, $J_0(w_0 r/R)$: despite the well-known oscillatory decreasing behavior of this function, because of the numerical values here at issue, the radial part can be approximated, in this specific example, with a simple second degree polynomial which decreases uniformly in the interval $0 \leq r \leq R$.

3.6 Conclusion

A number of cases have been discussed in which the complex world of thermal heat flux in various conditions can be reduced to a relatively simple mathematical form: the connection with experimental observations can be made quite immediate and effective if a thermal infrared imaging approach is adopted. Some special, otherwise not-so-much considered situations have been addressed: non-stationary configurations, mixed conductive/convective processes, non-uniform temperature fields and a somewhat unusual parallelism between electric and thermal laws all constitute a rich source of inspiration for conducting simple, visually enhanced experiments and to maintain a strong connection with theoretical models of intriguing appeal. We would also like to point out the opportunity of considering these classes of case studies in the framework of purely numerical simulations based on specific software platforms and apps. We mention, among others, the Energy2D simulation environment (Xie, 2012) which, with high interactivity allows the construction of even quite complex thermal configurations with conductive, convective and radiative contributions included. This free software can thus be used to create a visual and very direct counterpart to the experiments performed with the infrared thermal camera.

References

Çengel, Y. A. (2003). *Heat transfer: A practical approach* (2nd ed.). McGraw-Hill.
Crank, J. (1975). *The mathematics of diffusion* (2nd ed.). Clarendon.
Higgins, T. J. (1956). Electro-analogic methods. *Applied Mechanics Reviews, 1*, 1–12.
Incropera, F. P., & DeWitt, D. P. (1990). *Fundamentals of heat and mass transfer* (3rd ed.). Wiley.
Kirchhoff, G. (1845). Flow of an electric current through a plane, particularly through a circular disc. *Annalen der Physik und Chemie, 640*, 499–510.
Kohlrausch, F. (1884). On the guiding theory of thermoelectricity. *Annalen der Physik, 259*, 477–481.
Lienhard, J. H., IV, & Lienhard, J. H. V. (2019). *A heat transfer textbook* (5th ed.). Dover Pub.
Oss, S. (2020a). Infrared imaging of a non-stationary thermal conductive process and observation of its Green's kernel. *European Journal of Physics, 41*, 015102.
Oss, S. (2020b). Infrared imaging of the cooling fin equation. *European Journal of Physics, 41*, 055102.
Oss, S. (2021). Infrared visualization of lumped and non-lumped thermal transient processes in an introductory laboratory. *European Journal of Physics, 42*, 015101.
Oss, S., & Zendri, G. (2015). Electro-thermal analogies and imaging radiation thermometry. *European Journal of Physics, 36*, 065016.
Robertson, A. F., & Gross, D. (1958). An electrical-analog method for transient heat-flow analysis. *Journal of Research of the National Institute of Standards and Technology, 61*, 105–115.
Vollmer, M., & Möllmann, K.-P. (2018). *Infrared thermal imaging: Fundamentals, research and applications* (2nd ed.). Wiley-VCH.
Xie, C. (2012). Interactive heat transfer simulations for everyone. *The Physics Teacher, 50*, 237–240.

Part II
Science Education Research on the Use of Infrared Cameras

Chapter 4
Research on Educational Use of Thermal Cameras in Science: A Review

Christopher Robin Samuelsson

4.1 Introduction

The potential in using thermal or infrared (IR) cameras for educational purposes has been investigated in multiple studies during the twenty-first century. It is, however, difficult to say when that kind of research begun: The range of publications on thermal cameras goes from the types that describe ways of using the cameras for measurement without the purpose of using it for education (e.g. Jones, 2004; Pauk et al., 2019) to explicit educational use (experiments and phenomena) (e.g. Möllmann & Vollmer, 2007; Xie, 2011; Jeppsson et al., 2017) to empirical and theoretical papers on the educational use of thermal cameras (e.g. Haglund et al., 2015b; Larsson et al., 2019).

Beginning in the early 2000s, researchers (Möllmann & Vollmer, 2000; Vollmer, 2004; Vollmer et al., 2001) suggested experiments and phenomena that could be explored with thermal cameras. Around 2010, a second wave of instructive texts were published (Vollmer & Möllmann, 2010; Xie, 2011; Xie & Hazzard, 2011) in addition to the first international empirical study (Atkins et al., 2009).

Some of the experiments suggested in the instructive publications have later been applied as a basis for empirical studies involving students at different education levels investigating the experiments with IR cameras (e.g. Haglund et al., 2015b; Samuelsson et al., 2019b) and researchers have at times published one paper on the instructions and another where the instructions are tested with students (e.g. Haglund et al., 2017; Melander et al., 2016). This chapter will, however, mainly focus on the publications, on the use of IR cameras in science education, that are supported with data, i.e. *empirical publications*, in particular publications that apply

---

C. R. Samuelsson (✉)
Department of Physics and Astronomy, Uppsala University, Uppsala, Sweden
e-mail: robin.samuelsson@physics.uu.se

J. Haglund et al. (eds.), *Thermal Cameras in Science Education*, Innovations in Science Education and Technology 26, https://doi.org/10.1007/978-3-030-85288-7_4

some theoretical framework in analyzing and discussing the data. Many of these studies use "authentic" data in the sense that data is collected from natural settings, e.g. classrooms and labs. A total of *16 empirical publications* have been published to date.[1] Most of these studies involve experiments or activities on heat transfer[2] often relating temperature to heat transfer.

4.2 Research on Educational Use of IR Cameras: An Overview

Empirical publications involving IR cameras in education include participants from compulsory school to higher education in addition to public or informal education of visitors at museums or science centers (Atkins et al., 2009; Larsson et al., 2019) and householders (Boomsma et al., 2016; Goodhew et al., 2015). An overview of the *level of education* of the participants in each publication is given in Table 4.1. Higher education includes undergraduate and graduate level of the participants, compulsory school includes grade 1–9, upper secondary school refers to high school, informal education refers to education directed towards a general public audience but also school students visiting a science center and finally, multiple publications include more than one of the levels or type of education of the other categories and is in that case categorized as "*Multiple levels*".[3]

The participants have been engaged in specific subjects such as physics and chemistry or a more general type of science education. Most instructive publications deal with experiments aimed at teaching some physics or chemistry but there are also publications aimed at biology education. However, an explicit focus on technology education in publications on educational use of IR cameras is completely missing.

Table 4.1 Number of publications 1) on IR cameras in science education, involving a specific level of education of participants, and 2) on IR cameras in science education, involving a specific discipline of the course or laboratory work that the participants participate within

1) Level of education (number)	Higher education (5)	High school (1)	Compulsory school (4)	Informal education (4)	Multiple levels (2)
2) Discipline (number)	Physics (5)	Science (7)	Chemistry (1)	Biology (1)	Multiple disciplines (2)

[1] Publications written in languages other than English have been excluded. There are additional publications in review that may get published later this year (2020).

[2] Or the lack of heat transfer as in adiabatic expansion.

[3] The two publications that are categorized as "Multiple levels" include both high school and compulsory school, and compulsory school, higher education, high school and informal education.

An overview of the *disciplines* that the participants in each empirical study participate in, through their course or laboratory work, can also be found in Table 4.1. The participants and the content may be considered belonging to another discipline but it is the course or course unit that is referred to when describing the discipline (e.g. primary school teachers taking a physics course would be physics, physics students in a introductory chemistry class would be chemistry). It is possible that one publication includes more than one discipline. The publication is categorized as *"Multiple disciplines"* if that is the case.[4]

A majority of the empirical publications use a *qualitative methodology* involving video recording, transcription and *Prediction-Observation-Explanation* (White & Gunstone, 1992). Additionally, it seems that studies often are carried out in a *naturalistic* or authentic environment, i.e. the classrooms or labs where the students are regularly taught or at the museums and science centers where visitors would normally visit.

Two frameworks seem to dominate on the theoretical side: *Social semiotics* or Multimodality (Airey & Linder, 2017; Halliday, 1978; Kress & Hodge, 1988; Lemke, 1990) and the *Resources framework* (Hammer, 2000; Hammer et al., 2004; Redish, 2004) but other frameworks have also been applied in the research on IR cameras in education, for example *Conceptual metaphor theory* (Lakoff & Johnson, 1980) or theoretical constructs from other frameworks, for example *cognitive conflict* (Posner et al., 1982).

I will give an overview of the two dominating frameworks and the research on IR cameras in education that apply these frameworks to then move on to less frequently used frameworks or theoretical constructs.

4.3 Multimodality and Social Semiotics: An Overview

Multimodality is an approach to study the variety of means that meaning can be communicated with, such as colors (Kress & van Leeuwen, 2002), speech and gestures, at times referred to as modes or modalities (Jewitt et al., 2016). It is, however, difficult to talk about multimodality without "making explicit one's theoretical and methodological stance" (Jewitt et al., 2016, p. 1). Multimodality can thus be thought of as an umbrella term for frameworks and methods that include the study of additional modes of communication other than speech or written text. This includes methods and frameworks such as Conversation analysis, Systemic functional linguistics and Social semiotics, of which the last framework, including a more loose interpretation of multimodality, seems to have had most influence over the research on IR cameras in education.

[4] The two publications that are categorized as "Multiple disciplines" include both physics and science, and physics and chemistry.

Social semiotics can be said to have its origin in the work of Halliday (1978) and is about "the social dimensions of meaning, its production, interpretation and circulation" (Jewitt et al., 2016, p. 58). Examples of a social semiotic analysis include how a broad structure such as a culture or a more fine-grained structure such as a group dynamic affect the interpretation of a sign such as an emoji or a color such as red. The tools, for example the sign and the color in the previous example, used to derive meaning are sometimes referred to as *semiotic resources* (Jewitt et al., 2016).

Basically, social semiotics is, in contrast to traditional semiotics, contextualized; it is about the meaning made within a culture, an activity or a discipline, or in other words how one makes sense of what, for example, a sign, color or gesture means is affected by some social structures like nationality or type of learning activity. However, different accounts or versions of this framework have emerged over the years, adapted to the needs of the researchers. An early account building on Halliday's ideas was offered by Hodge and Kress (1988) in which they challenged the view of texts' meaning being static, i.e. that some meaning is inherent in the object or sign. As an example, let us have a look at the following exchange between a teacher and a student as presented by Lemke (1990, p. 20):

> "Teacher: [. . .] the white would be . . . ? Mark?"
> "Mark: Two S."

It may be difficult, if not impossible, to know what they are talking about (the meaning of "white" and "Two S") without knowing the context or social situation. In this case, the teacher is asking the student (Mark) about a part of an atomic representation colored white that represents an orbital (2 s) and not just the number 2 and the letter S or an abbreviation of "two seconds", in other words the meaning of "2 s" is not fixed but depends on the situation. Thus, the *affordance* of 2 s is different than in another situation not involving atomic orbitals and an exchange between a science teacher and a student.

The concept of affordance was offered by Gibson (1979) to in a broad sense capture the possible interactions between an organism and an object, which for the purposes of science education research could be a student and a coordinate system, or a teacher and a white board. Affordance "is what we normally pay attention to" (Gibson, 1979, p. 134) and can be thought of as the potential of an object when interacted with by a group or an individual: While a small child may use a chair as a ladder to reach a higher point to climb to, an adult would use the chair to sit on, i.e. the chair may have different affordances for small children and adults. However, the affordance depends on the context as the child may sit on the chair when it is dinner time and the adult may climb the chair when hammering a nail high up on the wall. Further elaboration of the concept of affordance has been made in Physics Education Research (Eriksson et al., 2014; Fredlund et al., 2012, 2014; Linder, 2013) where two types of affordance have been defined: *disciplinary affordance* (the potential of the object, or mode of communication, for practictioners in a discipline) and *pedagogical affordance* (the potential of the object, or mode of communication, in unpacking knowledge for learners). Essentially, these concepts describe if there is a potential use of the object for members of a discipline (discplinary affordance) and if there is a use in teaching content or a practice from that discipline (pedagogical affordance).

A discipline-based account of social semiotics has been developed by Airey and Linder (2009, 2017) in addition to Fredlund et al. (2015). In this approach to social semiotics, the meaning of, for example a sign, is set within the community of the discipline. However, this meaning does still depend on the context as one sign may have a range of meanings even within a specific discipline: Airey and Linder (2017) give the example of the so-called *right-hand rule* in physics that conveys how direction of an electric current relates to an induced magnetic field in a solenoid. Additionally, the right-hand rule can also be used to relate the direction of rotation for a plane to the angular momentum and angular velocity of a rotating wheel in a gyroscope. In this approach of social semiotics, the interpreter of a sign is supposed to appreciate it in a similar way as an experienced member in the discipline would, i.e. there is a practice to be learnt, e.g. to know when to interpret the right-hand rule as being about current or as being about direction of rotation. Additionally, this approach expands the concept of semiotic resource to also include specific laboratory apparatus (Airey & Linder, 2009, 2017), for example the IR camera.

4.4 Multimodality and Social Semiotics in Research on Educational Use of IR Cameras

Affordances of IR cameras have been investigated in several papers: For students of higher education levels (secondary school and undergraduate students) IR cameras have a *disciplinary affordance* related to macroscopic ideas in thermodynamics (Haglund et al., 2015a, b), which allows for adding extra dimensions of learning opportunity in phenomena that may have been missed without the cameras, for example the thermal properties of windows (Haglund et al., 2017). For younger students (secondary school, grade 7–8), Dolo et al. (2018) show that the IR cameras have a high *pedagogical affordance* that allows the students to integrate the *visual information* from the cameras with the *tactile information* from the hand and may lead to *true dialogue*[5]among the learners. This does, however, require that they are provided with some kind of simple model of the investigated phenomenon, i.e. that they have sufficient knowledge about the phenomenon before using the IR cameras. The presented papers make claims about the affordances of the IR camera *as a whole*. Other research (Samuelsson, 2020) on IR cameras has chosen to add another approach where not only the affordance of the camera is studied but also the affordances of some of the *aspects* of the IR camera (color, form and numbers) that are made meaningful in the learners' investigations with the cameras.

[5] A dialogue in which participants are "encouraged to pose their own questions, engage in collaborative inquiry, and develop complex lines of reasoning." (Dolo et al., 2018, p. 127)

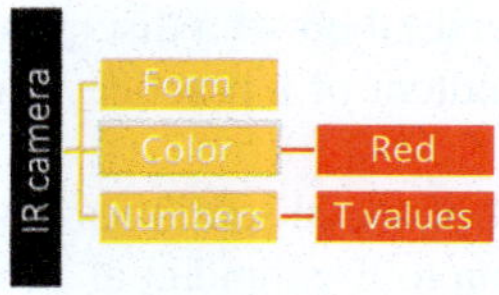

Fig. 4.1 The main aspects (yellow boxes) and sub-aspects (red boxes), of the IR cameras, used in the meaning making by participants investigating deliquescence and phase transitions. (Samuelsson, 2020)

In a study (Samuelsson et al., 2019a) on chemical engineering students', and their instructors', use of IR cameras in investigating deliquescence[6]it was concluded that the color scheme used (ranging from blue to red and white) seems to have a *pedagogical affordance*: Red and white were immediately associated to warmth by the students when observing the phenomenon with the cameras for the first time, which then led them to correctly conclude that the salt absorbed water from the surroundings, and that the salt exothermically reacted with the water. Additionally, the instructors used the cameras to advance their knowledge and challenge each other's explanations, which suggests a *disciplinary affordance* of the IR cameras (really the affordance of the semiotic resources coordinated in their investigation).

A closer analysis of the same data set, in comparison to an additional data set of teacher students using the IR cameras to investigate phase transitions, (Samuelsson, 2020) shows that three types of *aspects* and *sub-aspects*[7]are used in the meaning making of the participants (Fig. 4.1): the *form* of the cameras, the *colors* and the *numbers*. The numbers have the affordance of *measurement*, the form has the affordance of *spatial mobility*, one of the sub-aspects of the color, the color red, has the affordance of *attention to thermal aspects*, and finally the affordance of a sub-aspect of the numbers, measured temperature values, has the affordance of *attention to thermal aspects*. Different types of forms of the IR cameras show the same affordance, which is why it is the aspect, and not a sub-aspect, that has the affordance of spatial mobility.

4.5 The Resources Framework: An Overview

The *Resources framework* (e.g. Hammer, 2000; Hammer et al., 2004; Redish, 2004) was developed to emphasize contextuality in teaching and learning (as a response to research on conceptual change). The central concept in the framework is the concept

[6] A hygroscopic material (in this case sodium hydroxide) reacting with water from the surroundings.

[7] These so-called aspects and sub-aspects are, within the framework of social semiotics, called *semiotic systems* and *semiotic resources*, respectively.

of *resource*, which is an umbrella term for different ways and tools of knowing that may be incorrect in one context but correct, or *productive*, in another context. Such a resource may be a *phenomenological primitive*[8](p-prim) (DiSessa, 1993), a memory or some other cognitive element.

A set of *locally coherent resources* form the basis of *framing* (the answer to what a situation is about), a concept borrowed from anthropology (Goffman, 1986). As an encounter, or situation, is *framed*, a certain structure for *associations* is established in the situation that answers what the situation is about, or what to do in the situation. As an example: If one frames a situation to be about synthesizing a chemical compound one may have employed a set of resources involving one's *prototype*[9] of best lab practice, memories of synthesizing said compound and some p-prims or *heuristics* (shortcut reasoning strategies) to explain the involved chemical reactions.

Later accounts of the concept include different types of framing, for example *positional framing* (what roles one assigns to each participant) or *epistemological framing* (the kinds of knowledge one thinks is proper to use in a certain situation) (Sande & Greeno, 2012).

4.6 The Resources Framework in Research on Educational Use of IR Cameras

Studies in the field of IR cameras in education have identified many different resources that are productive in understanding or reasoning about thermal phenomena when using IR cameras for different contexts. Jeppsson et al. (2017) found that first-graders (primary school) reason about heat as a process when investigating thermal phenomena with IR cameras, and that they make heavy use of their everyday experience as *productive resources*. For example in metaphorically relating their previous experience to the scientific content of the encountered experiments (e.g. the water with the highest temperature was referred to as "hot as lava"). In a study (Schönborn et al., 2014) with older students (secondary school) using IR cameras to investigate thermal conductivity of different materials, the students did not have the resources that would have been productive in interpreting the encountered thermal phenomena. This may seem surprising as both studies probed the students' investigations with the same method (*Prediction-Observation-Explanation* (White and Gunstone, 1992)) and the participants in the latter study were older and thus more experienced with science than the students in the study of Jeppsson et al. (2017).

[8] A type of cognitive axiom or "basic statement about the functioning of the physical world that a user considers obvious and irreducible" (Redish, 2004, p. 57).

[9] A general category representation based on the average of one's experiences stored as some type of memory representation (Smith, 2014).

Another study (Samuelsson et al., 2019b) follows the result of everyday experience as *productive in reasoning* about thermal phenomena (Jeppsson et al., 2017). Primary school teacher students, supported with IR cameras and presented with tasks investigating and discussing thermal phenomena, contextualized in everyday situations, were encouraged to employ everyday resources as productive resources in finding a *coherence* across several contexts of similar science content (in this case phase transitions), i.e. to *transfer* the knowledge across different situations or contexts.

Other publications build on the concept of resource to instead, or additionally, study students' *framing* when using IR cameras:

Haglund et al. (2015b) found that some upper secondary school students tend to *conceptually frame* activities on thermal phenomena in similar ways (ignoring heat conduction in an activity on friction and using a macroscopic perspective in approaching an activity where a steel ball is dropped to the ground resulting in heat transfer and a temperature increase in the ground). They also found that the students *epistemologically framed* the activities in different ways, e.g. one student wanted to follow instructions while another wanted to deviate from the instructions.

In a thesis (Samuelsson, 2020) on students' (and instructors') reasoning when investigating thermal phenomena with IR cameras, it is suggested that engineering students *epistemologically frame* their investigation of a thermal phenomenon as "instructive" (follow instructions) while both primary school teacher students and the instructors of the engineering students epistemologically frame investigations of thermal phenomena as "inquiry" (exploration and testing). However, being more experienced with the practice of investigating scientific phenomena, the instructors structured their inquiry in ways that the primary school teacher students did not. The lack of structure in the primary school teacher students' practice, i.e. their *epistemological framing* as *unstructured inquiry*, led to several employments of resources that hindered or distracted their reasoning, so-called *barriers*.

In a study (Atkins et al., 2009) on how visitors at a public museum exhibit use the IR cameras, it was found that the visitors tend to frame the activities as lessons. However, the visitors are also able to *break out of this frame* by, for example, placing their own objects in front of the cameras or engage in discussions on topics not emphasized by the exhibition. The authors refer to this type of exhibit as a *hybrid activity space* (different types of activities are mixed and alternated between instructed and inquiry).

4.7 Other Learning Perspectives and Theoretical Constructs in Research on Educational Use of IR Cameras

In addition to the explicitly theory-based results presented above, there are also results more loosely based in traditional learning perspectives or results that are more difficult to tie to a specific framework. For example, multiple publications show how IR cameras invite to *instant inquiry* (e.g. Haglund et al., 2015a, 2016a, b), spontaneously posing new questions that can immediately be investigated with the IR cameras, for example whether water is cooled down or heated by blowing on it (Haglund et al., 2016b) or that it is possible to generate a *cognitive conflict* with the IR cameras (Larsson et al., 2019; Schönborn et al., 2014).

Research (Dolo et al., 2018; Haglund et al., 2015b, 2016a; Larsson et al., 2019; Schönborn et al., 2014) also indicates that the introduction of some kind of *simple model of heat transfer* is required to enable younger students to interpret the thermal images. Such a model could for example involve that heat transfer tends to occur between warm and cold objects when in contact and that insulators hinder the transfer (Haglund et al., 2015a). Older students (upper secondary school and higher education) may, however, not need as much support when interpreting thermal images generated by the cameras (Haglund et al., 2015a, b, 2016a, 2017).

Other studies indicate that IR cameras promote discussions about heat, temperature and related concepts across education levels and formal and informal education (Atkins et al., 2009; Dolo et al., 2018; Haglund et al., 2017; Jeppsson et al., 2017). Given sufficient knowledge about the phenomena before they are investigated with the cameras, students may engage in a *true dialogue* (Lemke, 1990) where they collaborate together to find answers and come up with their own questions. However, the types of talk engaged in when using the camera may differ between different students depending on their *epistemological framing* (as described earlier): Although the IR cameras may seem to promote exploration and inquiry in one context, students in another context may still resort to a repetitive and confirmative pattern of talk (so-called *cumulative talk* (Mercer, 1995)) when observing and explaining a thermal phenomenon with the camera (Samuelsson et al., 2019a), possibly as a result of the epistemological framing of the activity (Samuelsson, 2020). In addition, without sufficient knowledge about the phenomenon, students may rely on a *triadic dialogue* (Lemke, 1990) (teacher's question, student's answer and evaluation) rather than *true dialogue* (pose own questions and collaborate to find answers) (Dolo et al., 2018).

The IR cameras have also been shown (Larsson et al., 2019; Schönborn et al., 2014) to have the potential for generating *cognitive conflict* (Posner et al., 1982) – a theoretical construct describing the discomfort experienced by, for example a student, when encountering some information that goes against one's preconceptions or knowledge prior to the encounter. This may lead to a conceptual change (analogous to the conceptual change of a discipline or science (Kuhn, 2012)) for the student wherein one accommodate the new knowledge. Conceptual change has, however, been shown to be very difficult to achieve as students who apparently have showed signs of conceptual change may still retain their so-called naïve theories (Dunbar et al., 2007).

In a study by Schönborn et al. (2014) it was found that secondary school students (7th grade) observing that a metallic knife has almost the same temperature as a piece of wood even though the knife feels much colder than the wood were bewildered by this anomaly but could not resolve the cognitive conflict. In another study (Larsson et al., 2019), applying a *metaphor analysis*, the observations with the

IR cameras, in addition to the sensory experience, also lead to a cognitive conflict but this resulted in the students distrusting the information conveyed by the cameras rather than their sensory experience. A similar result was also found in a study by Haglund et al. (2015b), where the students decided to explain the conflicting observations with the IR camera acting as an "other-way-around machine" (Haglund et al., 2015b, p. 15). Larsson et al. (2019) conclude that what happens in their study is that the students lose "access to any heat metaphors and are unable to reach any conclusion" (Larsson et al., 2019, p. 85). The same study also concludes that the colors displayed by the cameras are used as both metonyms and metaphors for heat by the students: The students tended to use the color as a metonym when "heat" was conceptualized as a static entity and as a metaphor when "heat" was conceptualized as dynamic. Additionally, they almost exclusively used "heat" as a noun. An earlier study (Brookes & Etkina, 2015) on students' reasoning about heat has shown that the way students talk about heat tells us about how they understand heat as either a process, quantity or entity (represented by either a caloric, operational or process definition of heat).

Some papers explore potential teaching sequences, with a set amount of steps involving tasks both with and without access to IR cameras: Nordine and Wessnigk (2016) designed a sequence for middle school students, which, when tested with some students, indicated that IR cameras can support the understanding that transfer of thermal energy can account for energy that appears to go missing in a phenomenon. In line with previous ideas (e.g. Schönborn et al., 2014) on students having sufficient knowledge about a phenomenon before exploring it with IR cameras, the students were to first learn about and discuss the content of the phenomena later investigated with the IR cameras. A second step included the introduction of the IR cameras as a task itself, e.g. the students get to explore the environment with a set of prompts and reflective questions for support. This then transitioned into the investigation of phenomena involving energy that goes "missing" without IR cameras, like when sliding objects of different materials down a slope. A final reflection of students participating in this sequence included comments on how they appreciated using the cameras and that they learnt that "heat" does not vanish but rather spreads out until everything has an equal temperature.

Samuelsson et al. (2019b) developed a teaching sequence, involving investigations of thermal phenomena with IR cameras, in which students' everyday knowledge was leveraged in teaching about phase transitions and energy transfer. The teaching sequence involved a simple model of energy transfer related to phase transitions and three parts framed by different contexts: condensation on your body in a sauna, evaporation from your body in a shower and evaporation from a glass of water left on a table. The students were to discuss the first two situations based on experienced sensations and got to investigate experiments, related to each situation, with IR cameras, for example sprinkling some water on an arm. The results showed that the students managed to relate each part to the others (really the experiments involving IR cameras to the subsequent and previous everyday situations) to find a *coherence* across all contexts and that they often turned to the second law of thermodynamics and previous experiences when attempting to explain or understand the investigated phenomena.

Unger et al. (2020) designed and tested a teaching sequence on endothermic and ectothermic animals. The sequence begins with a presentation and overview of the content of the sequence, including the concepts of endothermic, ectothermic and thermoregulation in addition to some questions on the taught content for the students to answer. The basis for the sequence thus aligns with previous papers recommending or showing that students using IR cameras should be supported with some model or knowledge about the phenomena they are to investigate with IR cameras (e.g. Dolo et al., 2018; Schönborn et al., 2014). The students are then to investigate the temperatures of an endothermic animal (a mouse) and ectothermic animals (lizards) with IR cameras. Data is later analyzed, compared and discussed together with written questions to answer. The study shows that a large majority (86%) of the students agreed or strongly agreed that the IR cameras supported an understanding of the difference between endothermic and ectothermic animals.

Additionally, studies (Boomsma et al., 2016; Goodhew et al., 2015) on public learning of energy conservation indicate that the thermal imaging of IR cameras can change the long-term behavior of individuals: Goodhew et al. (2015) provided one group of householders thermal images of their homes, one group with carbon footprint audits and one with nothing (the two latter acted as control groups). The householders that had viewed thermal images of their homes had reduced their energy use when revisited a year later. Another data set with householders showed that they were about 5 times more likely to install draught proofing measures, especially addressing issues visible in the thermal images, in their houses after having viewed thermal images of their homes. A later study (Boomsma et al., 2016) showed that tailoring the thermal images to be more relevant or personal to the householder viewing them lead to more engagement in regards to sharing the images and intents of going farther in energy savings than others having watched non-tailored images. It was also shown that when comparing the *recall* of the householders having watched the thermal images with householders having read a text, the thermal images lead to more *vividness* in the recall, i.e. they were more likely to excite the imagination and hold the *attention* of the viewers.

4.8 Concluding Overview

It is common to study the use of IR cameras in education through *qualitative* methods rather than *quantitative* (and data is often collected through video recording which is then transcribed). A preferred method for probing the students about their reasoning is *Prediction-Observation-Explanation* (White & Gunstone, 1992).

Multimodal theory like *Social semiotics* in addition to the *Resources framework* seem to be the more common theoretical frameworks applied to studies on the use of IR cameras in science education. Theoretical constructs applied from other frameworks include *vividness*, *cognitive conflict* and *conceptual metaphors*.

The use of IR cameras seems to relate to the *education level* and *subject knowledge* of the users: younger students need support in using the cameras by, for example, providing them with a simple model of the investigated phenomenon. Additionally, insufficient knowledge about the phenomenon may lead to triadic dialogue when using the IR cameras. It is recommended that IR cameras be used together with some *supporting model* or only when the students have sufficient knowledge about the subject content involved in the phenomenon to be investigated with the IR cameras.

Students' *framing* of tasks or teaching sequences involving IR cameras may differ depending on previous experiences and the type of education they participate in. The use of IR cameras may contribute to a *coherent framing* across contexts.

Everyday experiences, such as the sense of touch, can act as *productive resources* when investigating and discussing thermal phenomena with IR cameras either to *link intuition to disciplinary content* or in creating a *cognitive conflict*. However, their previous experiences may also lead to confusion (unresolved cognitive conflicts) if the learners are unable to relate them to the disciplinary content and therefore choose to rely more on their experiences than the cameras. In other words, the IR cameras can *mediate* between everyday experience and the disciplinary content but this requires that the learners are either experienced enough to understand the investigated phenomenon or presented with some teaching material or teaching that can support the use of the IR cameras, i.e. supported through instructional scaffolding. The *colors*, and especially the color red, seem to be important for this mediation through, for example, the use of *metaphors* and metonyms involving the colors.

Furthermore, the colors of the thermal images are important in focusing the *attention* of the users, as is the *vividness* of the thermal image as a whole. Watching thermal images may also lead to a better recall than reading a text with similar information. Additionally, personal or *tailored* thermal images may lead to a stronger engagement with the images.

The research on the educational use of IR cameras begun in the early twenty-first century and has up until now shown that there is an educational potential in using the IR cameras but that students' investigations with the IR cameras often need to be supported through instructional scaffolding. However, there are still many unanswered questions about the educational use of IR cameras in science education: How do students' investigations with the IR cameras vary with the instructional method? How can IR cameras be used in technology education? What are the long-term effects on students' learning in thermal science when using IR cameras in lab work? These and other questions could be answered in future publications on IR cameras in science education.

References

Airey, J., & Linder, C. (2009). A disciplinary discourse perspective on university science learning: Achieving fluency in a critical constellation of modes. *Journal of Research in Science Teaching, 46*(1), 27–49. https://doi.org/10.1002/tea.20265

Airey, J., & Linder, C. (2017). Social semiotics in university physics education. In D. F. Treagust, R. Duit, & H. E. Fischer (Eds.), *Multiple representations in physics education* (pp. 95–122). Springer.

Atkins, L. J., Velez, L., Goudy, D., & Dunbar, K. N. (2009). The unintended effects of interactive objects and labels in the science museum. *Science Education, 93*(1), 161–184.

Boomsma, C., Goodhew, J., Goodhew, S., & Pahl, S. (2016). Improving the visibility of energy use in home heating in England: Thermal images and the role of visual tailoring. *Energy Research and Social Science, 14*, 111–121.

Brookes, D. T., & Etkina, E. (2015). The importance of language in students' reasoning about heat in thermodynamic processes. *International Journal of Science Education, 37*(5–6), 659–779. https://doi.org/10.1080/09500693.2015.1025246

DiSessa, A. A. (1993). Toward an epistemology of physics. *Cognition and Instruction, 10*(2), 105–225.

Dolo, G., Haglund, J., & Schönborn, K. (2018). Thermal cameras as a semiotic resource for inquiry in a South African township school context. *Designs for Learning, 10*(1), 123–134.

Dunbar, K. N., Fugelsang, J. A., & Stein, C. (2007). Do naive theories ever go away? Using brain and behavior to understand changes in concepts. In M. Lovett & P. Shah (Eds.), *Carnegie Mellon symposia on cognition. Thinking with data* (pp. 193–205). Lawrence Erlbaum Associates Publishers.

Eriksson, U., Linder, C., Airey, J., & Redfors, A. (2014). Introducing the anatomy of disciplinary discernment: An example from astronomy. *European Journal of Science and Mathematics Education, 2*(3), 167–182.

Fredlund, T., Airey, J., & Linder, C. (2012). Exploring the role of physics representations: An illustrative example from students sharing knowledge about refraction. *European Journal of Physics, 33*(3), 657–666. https://doi.org/10.1088/0143-0807/33/3/657

Fredlund, T., Linder, C., Airey, J., & Linder, A. (2014). Unpacking physics representations: Towards an appreciation of disciplinary affordance. *Physical Review Special Topics - Physics Education Research, 10*(2), 1–13. https://doi.org/10.1103/PhysRevSTPER.10.020129

Fredlund, T., Linder, C., & Airey, J. (2015). A social semiotic approach to identifying critical aspects. *International Journal for Lesson and Learning Studies, 4*(3), 302–316.

Gibson, J. J. (1979). The ecological approach to visual perception. In J. J. Gibson (Ed.), *The theory of affordances* (pp. 127–143). Houghton Miffin.

Goffman, E. (1986). *Frame analysis: An essay on the organization of experience*. Northeastern Univ. Press.

Goodhew, J., Pahl, S., Auburn, T., & Goodhew, S. (2015). Making heat visible: Promoting energy conservation behaviors through thermal imaging. *Environment and Behavior, 47*(10), 1059–1088. https://doi.org/10.1177/0013916514546218

Haglund, J., Jeppsson, F., Hedberg, D., & Schönborn, K. J. (2015a). Thermal cameras in school laboratory activities. *Physics Education, 50*(4), 424–430.

Haglund, J., Jeppsson, F., Hedberg, D., & Schönborn, K. J. (2015b). Students' framing of laboratory exercises using infrared cameras. *Physical Review Special Topics – Physics Education Research, 11*(2), 1–22.

Haglund, J., Jeppsson, F., Melander, E., Pendrill, A.-M., Xie, C., & Schönborn, K. J. (2016a). Infrared cameras in science education. *Infrared Physics and Technology, 75*(March), 150–152. https://doi.org/10.1016/j.infrared.2015.12.009

Haglund, J., Jeppsson, F., & Schönborn, K. J. (2016b). Taking on the heat—A narrative account of how infrared cameras invite instant inquiry. *Research in Science Education, 46*(5), 685–713.

Haglund, J., Melander, E., Weiszflog, M., & Andersson, S. (2017). University physics students' ideas of thermal radiation expressed in open laboratory activities using infrared cameras. *Research in Science and Technological Education, 35*(3), 349–367. https://doi.org/10.1080/02635143.2017.1318362

Halliday, M. A. K. (1978). *Language as social semiotic: The social interpretation of language and meaning*. University Park Press.

Hammer, D. (2000). Student resources for learning introductory physics. *American Journal of Physics, 68*(S1), S52–S59. https://doi.org/10.1119/1.19520

Hammer, D., Elby, A., Scherr, R. E., & Redish, E. F. (2004). Resources, framing, and transfer. In J. Mestre (Ed.), *Transfer of learning: Research and perspectives* (pp. 1–26). Information Age Publishing.

Jeppsson, F., Frejd, J., & Lundmark, F. (2017). "Wow, it turned out red! First, a little yellow, and then red!" 1st-graders' work with an infrared camera. *Journal of Research in Childhood Education, 31*(4), 581–596. https://doi.org/10.1080/02568543.2017.1347589

Jewitt, C., Bezemer, J., & O'Halloran, K. (2016). *Introducing multimodality*. Routledge.

Jones, H. G. (2004). Application of thermal imaging and infrared sensing in plant physiology and ecophysiology. In J. A. Callow (Ed.), *Advances in botanical research* (Vol. 41, pp. 107–163). Elsevier Academic Press.

Kress, G., & Hodge, R. (1988). *Social semiotics*. Cornell University Press.

Kress, G., & van Leeuwen, T. (2002). Colour as a semiotic mode: Notes for a grammar of colour. *Visual Communication, 1*(3), 343–368. https://doi.org/10.1177/147035720200100306.

Kuhn, T. S. (2012). *The structure of scientific revolutions*. University of Chicago Press.

Lakoff, G., & Johnson, M. (1980). *Metaphors we live by*. University of Chicago Press.

Larsson, A., Stafstedt, M., & Schönborn, K. J. (2019). Heat angels and paper cups: Pupils' use of metaphoric relations when engaging thermal cameras to investigate heat. In R. Millar (Ed.), *Bridging research and practice in science education* (pp. 73–89). Springer Nature Switzerland.

Lemke, J. L. (1990). *Talking science: Language, learning, and values*. Ablex.

Linder, C. (2013). Disciplinary discourse, representation, and appresentation in the teaching and learning of science. *European Journal of Science and Mathematics Education, 1*(2), 43–49.

Melander, E., Haglund, J., Weiszflog, M., & Andersson, S. (2016). More than meets the eye – Infrared cameras in open-ended university thermodynamics labs. *The Physics Teacher, 54*(9), 528–531.

Mercer, N. (1995). *The guided construction of knowledge: Talk among teachers and learners*. Multilingual Matters.

Möllmann, K.-P., & Vollmer, M. (2000). Eine etwas andere, physikalische Sehweise. *Physikalische Blätter, 56*(9), 65–69.

Möllmann, K.-P., & Vollmer, M. (2007). Infrared thermal imaging as a tool in university physics education. *European Journal of Physics, 28*(3), S37–S50. https://doi.org/10.1088/0143-0807/28/3/S04

Nordine, J., & Wessnigk, S. (2016). Exposing hidden energy transfer with inexpensive thermal imaging cameras. *Science Scope, 39*(7), 25–32.

Pauk, J., Wasilewska, A., & Ihnatouski, M. (2019). Infrared thermography sensor for disease activity detection in rheumatoid arthritis patients. *Sensors, 19*(16), 1–14.

Posner, G., Strike, K., Hewson, P., & Gertzog, W. (1982). Accommodation of a scientific conception: Toward a theory of conceptual change. *Science Education, 66*(2), 211–227. https://doi.org/10.1002/sce.3730660207

Redish, E. (2004). A theoretical framework for physics education research: Modeling student thinking. In *Proceedings of the International School of Physics "Enrico Fermi" Course CLVI: Research on Physics Education* (pp. 1–56).

Samuelsson, R. (2020). *Reasoning with thermal cameras settings in higher education*. Uppsala University.

Samuelsson, C. R., Elmgren, M., & Haglund, J. (2019a). Hot vision: Affordances of infrared cameras in investigating thermal phenomena. *Designs for Learning, 11*(1), 1–15. https://doi.org/10.16993/dfl.94

Samuelsson, C. R., Elmgren, M., Xie, C., & Haglund, J. (2019b). Going through a phase: Infrared cameras in a teaching sequence on evaporation and condensation. *American Journal of Physics, 87*(7), 577–582. https://doi.org/10.1119/1.5110665

Schönborn, K., Haglund, J., & Xie, C. (2014). Pupils' early explorations of thermoimaging to interpret heat and temperature. *Journal of Baltic Science Education, 13*(1), 118–132.

Smith, J. D. (2014). Prototypes, exemplars, and the natural history of categorization. *Psychonomic Bulletin and Review, 21*, 312–331. https://doi.org/10.3758/s13423-013-0506-0

Unger, S. D., Rollins, M. A., & Thompson, C. M. (2020). Hot- or cold-blooded? A laboratory activity that uses accessible technology to investigate thermoregulation in animals. *The American Biology Teacher, 82*(4), 227–233.

Van De Sande, C. C., & Greeno, J. G. (2012). Achieving alignment of perspectival framings in problem-solving discourse. *Journal of the Learning Sciences, 21*(1), 1–44. https://doi.org/10.1080/10508406.2011.639000

Vollmer, M. (2004). Physics of the microwave oven. *Physics Education, 39*(74), 74–81.

Vollmer, M., & Möllmann, K.-P. (2010). *Infrared thermal imaging: Fundamentals, research and applications*. Wiley-VCH.

Vollmer, M., Möllmann, K.-P., Pinno, F., & Karstädt, D. (2001). There is more to see than eyes can detect. *The Physics Teacher, 39*(6), 371–376. https://doi.org/10.1119/1.1407135

White, R., & Gunstone, R. (1992). *Probing understanding*. Falmer Press.

Xie, C. (2011). Visualizing chemistry with infrared imaging. *Journal of Chemical Education, 88*(7), 881–885. https://doi.org/10.1021/ed1009656

Xie, C., & Hazzard, E. (2011). Infrared imaging for inquiry-based learning. *The Physics Teacher, 49*(6), 368–372. https://doi.org/10.1119/1.3628268

Chapter 5
Using Thermal Cameras in Secondary Physics to Support Learning About Energy

Jeffrey Nordine, Susanne Weßnigk, and Larissa Greinert

5.1 Introduction

Conservation is central to the nature of energy, yet energy is certainly not the only conserved quantity – others include: charge, linear momentum, and angular momentum. What is remarkable about energy is the incredibly wide range of phenomena to which the energy concept can be applied. For example, while conservation of linear momentum is primarily useful for making sense of phenomena when relatively large forces act over relative short time (e.g., in collisions), conservation of energy is useful for making sense of an incredibly diverse set of phenomena, including: population dynamics, climate change, electricity, chemical reactions, and on and on. As a conserved quantity, energy is special in terms of the breadth of its usefulness across the sciences and engineering. The widespread applicability of the energy concept has contributed to its ubiquity in both science and everyday life. Well before students encounter the energy concept in school instruction, students use energy ideas and terms to describe why they need to eat food, the need to charge a cell phone, etc. In everyday settings, students typically observed energy as a quantity that is "used up" and needs to be replenished within living systems and technical devices (Driver et al., 1994; Trumper, 1993; Watts, 1983). Put simply, students' everyday experiences overwhelmingly conflict with the idea that energy is conserved.

J. Nordine (✉)
Leibnitz Institute for Science and Mathematics Education (IPN), Kiel, Germany
e-mail: nordine@leibniz-ipn.de

S. Weßnigk
QUEST-Leibniz-Forschungsschule, Leibniz University Hannover, Hannover, Germany
e-mail: susanne.wessnigk@quest.uni-hannover.de

L. Greinert
Ricarda-Huch-Schule Hannover, Hannover, Germany
e-mail: larissa.greinert@rhshannover.de

J. Haglund et al. (eds.), *Thermal Cameras in Science Education*, Innovations in Science Education and Technology 26, https://doi.org/10.1007/978-3-030-85288-7_5

With respect to energy conservation, students' everyday experiences typically seem to contradict the ideas that they are taught in the course of formal science instruction (Duit, 2014; Solomon, 1983). Despite a lifetime of experiences noting that the energy originally stored within a system decreases over time, they are explicitly taught in physics class to assume that the energy of a system of interest remains the same over time. Yet, students can plainly see that a ball does not bounce as high as from where it was dropped and that a pendulum will not swing back to its release height. To explain this discrepancy between theory and observation, physics teachers quite rightly point out that energy has been transferred to the surroundings and/or transformed into thermal energy, yet this claim is seldom supported by direct observation.

Asking students to accept scientific principles without evidence is problematic on both cognitive and epistemological grounds. From a cognitive perspective, students construct understanding based upon their experiences with the world (National Academies of Sciences, Engineering, and Medicine, 2018; National Research Council, 2000), and new ideas replace or augment existing ideas only when they prove fruitful for making better sense of the world and better align with their observations (Duit & Treagust, 2003; Smith et al., 1993). From an epistemological perspective, asking students to accept a principle as foundational to science as energy conservation without empirical evidence is antithetical to the nature of scientific inquiry (Lederman, 2007).

Thermal cameras can provide students with critical evidence for the conservation of energy by making visible the tiny energy transfers that result in tiny, transient temperature changes that are simply not perceptible by traditional thermometers. In this way, thermal cameras can illuminate the discrepancies between idealized models of phenomena in which energy conservation within a system is assumed and real-world situations in which energy is transferred to the surroundings via mechanisms that are difficult to detect (Netzell et al., 2017).

Until relatively recently, using thermal cameras in the context of school science instruction has been prohibitively expensive. Newer hand-held devices cost about as much as a typical electronic probe and are designed to be used with a smartphone or tablet with little training. Yet, simply showing students images from a thermal camera in a whole-class demonstration or even putting a thermal camera directly in students' hands is unlikely to provide the learning experiences necessary for students to change their pre-existing ideas about energy or to support student engagement in meaningful inquiry using these powerful tools.

We argue that thermal cameras play an important role in engaging students in energy-related inquiry that supports their understanding of energy. In the coming sections, we first review research into student learning about energy and how student ideas develop through instruction, then we discuss implications for the use of thermal cameras in physics instruction in terms of curricular design and student supports. Finally, we synthesize empirical findings from empirical research illustrating how intentionally-designed instruction can leverage the benefits of thermal cameras to strengthen student learning about energy in secondary physics.

5.2 Student Learning About Energy

Students' development of intuitive and instructed ideas about energy has long been a focus within science education research. In the 1980s and 1990s, researchers explored and documented common student conceptions about energy that conflicted with scientific consensus. Michael Watts (1983) found that secondary students commonly held one or more of seven categories of non-normative ideas about energy: energy is only associated with living things, energy is associated with obvious activity, energy is held by some object/devices and needed by others, energy is a short-lived by-product of certain processes, energy is a dormant ingredient in some substances that is activated by a trigger, and energy is a functional fuel for powering technical devices, energy is a physical fluid that flows, e.g., in circuits. Watts' original categories were later substantiated by other researchers (Nordine et al., 2011; Trumper, 1993).

Since the original identification of common alternative ideas about energy, researchers have explored how students tend to develop energy ideas over time. Driver et al. (1994) synthesized research into student conceptions and proposed that students progress along a fairly common sequence, beginning with their own sense of feeling energetic, applying this sense to other living things and technical devices, recognizing that energy can be stored, and finally progressing to learning about energy conservation and degradation. In the past 15 years, a spate of large-scale, cross-sectional studies have endeavored to empirically construct learning progressions (Alonzo & Gotwals, 2012) that describe common pathways for student development of energy understanding. Liu and McKeough (2005) found that students begin by recognizing that energy is associated with various activities, has various sources/forms, can be transferred, is degraded and/or "lost" to the surroundings during most processes, and finally that energy is conserved. Importantly, Liu and McKeough reported that only a small minority of students develop an understanding of energy conservation by the end of secondary education. Neumann et al., (2013) similarly found that students' ideas progressed from recognizing that energy can exist in various forms, energy can be transferred/transformed, degraded/dissipated, and finally conserved, with few students demonstrating understanding of energy conservation. Herrmann-Abell and DeBoer (2018) reported a similar progression in student ideas and observed similar levels of difficulty in developing ideas about energy degradation and conservation. Overall, results from empirical study into how students develop energy ideas over time demonstrates that the majority of students begin by recognizing various manifestations of energy, then come to understanding energy as a quantity that can be transferred between systems and/of transformed between forms. From there, many students struggle with the idea that energy is degraded and/or dissipated in macroscopic phenomena and very few students develop robust and useful ideas about energy conservation.

Results of research into student ideas and learning about energy offer two key findings. First, around middle school, most students come to understand energy as a quantity that can be transferred between systems and/or transformed between its various manifestations (i.e., forms) as phenomena occur. Second, even at the end of secondary school, students struggle to understand and use the principle of energy conservation, perhaps because it is at odds with their intuitive ideas and experiences.

By providing direct evidence of the hidden energy transfers that often obscure energy conservation, thermal cameras have the potential for supporting students in moving beyond accepting that energy can be transferred/transformed to understanding energy as a conserved quantity that is often dissipated in everyday processes. In order to leverage this potential, instruction must be designed so that students are both technically and conceptually prepared to benefit from viewing the images produced a thermal camera.

5.3 Designing Effective Energy Instruction Using Thermal Cameras

Thermal cameras are a potentially transformative device for the teaching and learning of energy, yet, like any tool, the manner of its use will greatly influence whether and how students benefit from its inclusion in instruction. In this section, we argue that energy instruction that effectively leverages the potential of thermal cameras must account for students' prior knowledge, strategically introduce thermal cameras when students are ready to benefit from them, and include specific scaffolds for supporting students in using thermal cameras to engage in scientific inquiry.

5.3.1 The Role of Prior Knowledge in Interpreting Thermal Camera Images

Students learn based on what they already know, and this influences both what new ideas are accessible and how students interpret new information (Bransford & Schwartz, 1999). Accordingly, students will make sense of the images produced by thermal cameras in light of their prior knowledge. For example, students who believe that energy is a short-lived by-product of certain processes will likely have their ideas reinforced when using a thermal camera to observe that technical devices are warm when they operate and cool relatively quickly when they are turned off. The observation that people tend to be warmer than their surroundings may reinforce the idea that energy is primarily associated with living things. If not used carefully, thermal cameras can reinforce non-normative ideas about energy. Thus, instruction must set the stage for students to interpret thermal camera images such that they reinforce rather than undermine learners' acceptance of normative ideas about energy. To do this, instructional designers should carefully consider what ideas that learners should hold prior to introducing thermal cameras within the course of instruction. Accordingly, it is important to consider which ideas about energy thermal cameras are uniquely positioned to support.

Thermal cameras offer relatively little support for the idea that energy can be manifested in different ways (i.e., exist in different forms); this idea is better taught using naked-eye observations of phenomena, such as a ball falling, a spring bouncing, a hand-cranked electric generator powering a light bulb, etc. Likewise, thermal cameras are of relatively little use in providing evidence for the idea that energy can be transferred between systems. For this, students can observe collisions, directly feel temperature changes when a warm object is placed in thermal contact with a cooler object, or observe a Crooke's radiometer start spinning when a flashlight shines on it. Thermal cameras can expose hidden energy transfers, but teaching the idea that energy *can be transferred* is best done through overt interactions between systems that result in clearly observable changes.

Thermal cameras are uniquely well-positioned to provide evidence for energy dissipation, which is foundational to supporting students in accepting the idea of energy conservation. To recognize images from thermal cameras as evidence for these ideas, students must be prepared through a set of coherent instructional activities that set the stage for productive learning.

5.3.2 Strategically Embedding Thermal Cameras in Coherent Instruction

At their core, thermal cameras function as a measurement tool that supports inquiry into one's environment; science instruction that effectively leverages the potential of thermal cameras should recognize this core function of thermal cameras by embedding their use into broader scientific investigations in which students are engaged gathering evidence (i.e., data used to evaluate a hypothesis) to answer questions about their world. That is, thermal cameras are best positioned to support student learning when students use them as a tool for inquiry.

When instruction involves a set of connected learning activities in which students collaborate to make sense of relevant phenomena and problems over extended time, this is coherent instruction (Kali et al., 2008). Coherence is especially important in the teaching and learning of energy, as energy is a multifaceted and abstract science concept that takes many years to learn (Chen et al., 2014; Nordine, 2016). Students therefore need ample opportunities to develop and test ideas using evidence gathered from thermal cameras, and it is important to carefully define what prior knowledge about energy is expected as well as what energy-related competencies students should develop during a coherent instructional unit. That is, coherent instruction should align with research exploring how students build ideas about energy over time (Fortus & Krajcik, 2012). Further, coherent instruction should carefully sequence activities into a "storyline" that captures and sustains student interest by spurring an appropriate sense of puzzlement (i.e., new problems are accessible given students' existing knowledge) about how/why phenomena occur and motivating a need-to-know about new ideas before they are introduced (Nordine et al., 2019).

One particularly useful model for designing coherent instruction is project-based pedagogy (Krajcik & Czerniak, 2018; Schneider et al., 2020). In project-based pedagogy, learning is motivated and contextualized through a driving question that originates from students' initial explorations of relevant phenomena. Examples of driving questions in secondary school energy instruction include "How can I use trash to power my stereo?" (Nordine et al., 2011) and "Why do some things stop and others keep going?" (Fortus et al., 2013). These driving questions connect to students' experiences outside of school, require energy ideas to answer, and are rich enough to sustain a series of connected instructional activities. Coherent energy instruction focuses on answering questions about relevant phenomena, and thermal cameras play a central role in gathering data that is useful for answering those questions. For example, along the way to investigating the question "How can I use trash to power my stereo?", it is important to explore why it is necessary to "power" technical devices; such an exploration involves learning about energy transfers to and from a technical device. While energy transfers to a technical device are more obvious (e.g., an electrical outlet or a heat source), energy transfers from such a device are more difficult to detect. Thermal cameras play a key role in establishing that when devices are "powered", energy is transferred both to and from the device. This observation directly challenges the common conception that energy is "used up" by technical devices and provides evidence that energy is transferred to the environment in ways that are not obvious to our naked eye. Over time and across a set of investigations, students build a robust evidence base that supports the more general principles of energy conservation and dissipation. Importantly, thermal cameras are an authentic tool used to investigate real-world problems and phenomena (e.g., to identify insulation problems in windows); using such authentic tools is a core part of project-based science that helps to strengthen connections between school science learning and everyday life.

In accordance with their broad utility in everyday life and professional contexts, thermal cameras have become more widely available and inexpensive in recent years. The cost of thermal cameras has decreased to the point where it is possible to purchase several or more thermal cameras for a single school, which fundamentally changes the way that students can use cameras in the context of coherent instruction. Students can control how thermal cameras are used in their investigations and begin to make decisions about how to use these measurement tools to investigate phenomena, setting the stage for deeper engagement and shared sense-making in small groups. While student control over thermal cameras carries significant learning benefits, there are also challenges in supporting students in using thermal cameras effectively. Explicit student scaffolds are critical for supporting students in properly gathering and interpreting data from thermal cameras.

5.3.3 *Student Supports for Using Thermal Cameras*

Perception of visual images is highly influenced by the predictive mental models and relevant domain-specific knowledge held by the viewer (de Lange et al., 2018). Thermal cameras present a new challenge for learners as they interpret the visual images produced by them, since learners tend to have very little experience with non-visible radiation and may not readily accept the idea that all bodies emit thermal radiation (S. Neumann & Hopf, 2012). This paucity of experience and prior domain-specific knowledge provides fertile ground for students to struggle creating meaning about the visual images produced by a thermal camera. For example, students viewing thermal camera images may incorrectly interpret what they show, for example, attributing areas with a color corresponding to colder regions as "transmitting cold" (Haglund et al., 2015; Hoppe, 2018). Additional challenges in interpreting the images/videos produced by thermal cameras include: dynamic adjustments of temperature scale, thermal reflection and the emission coefficient of observed materials, thermal and geometric resolution, and conceptually linking the thermal image/video to the phenomenon under consideration (Greinert & Weßnigk, 2020). Students need specific supports for successfully using thermal cameras to gather and interpret evidence of thermal transfers in phenomena.

One promising approach to ensuring that students effectively use thermal cameras is the incorporation of a "Driver's License" (Greinert & Weßnigk, 2020) for using thermal cameras independently. In this approach, students are asked to view and interpret several videos taken by thermal cameras (prepared in advance by an instructor). Videos (see Fig. 5.1) are carefully constructed to elicit common difficulties, for example:

- focusing on a ball as a hand enters the video and leaves, causing the color of the ball to dynamically adjust despite constant temperature,
- showing a ball that is picked up and replaced by a hand, causing the temperature of the ball to increase, and

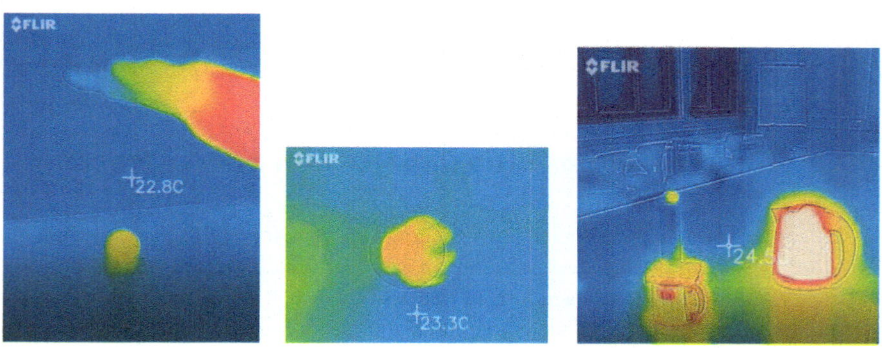

Fig. 5.1 Screen captures from example videos interpreted by students in order to attain their thermal camera Driver's License (Greinert, 2020)

- showing objects of different temperatures, then zooming the frame to focus on various objects and causing the color scaling of objects to dynamically adjust (Greinert & Weßnigk, 2020).

For each video, students work in small groups to discuss the action shown in each video and to interpret how the colors of the target object changes throughout the video, drawing conclusions about whether the target object changes temperature.

When students begin to record their own videos of phenomena, they are given an informational card intended to scaffold their creation of videos as they gain expertise. This informational card outlines key criteria for taking useful videos with a thermal camera, such as:

- ensuring that relevant temperature differences are visible
- exclusion of unnecessary objects or people, including residual "footprints" and "handprints" from recently-touched areas
- using appropriate zoom, camera work (e.g., static/dynamic framing), time of recording (Köhn, 2019).

The thermal camera Driver's License is an example of a scaffold that supports students in both the interpretation of images/videos produced by a thermal camera and the independent use of thermal cameras to capture images/videos that yield high-quality data about the phenomenon under investigation. These student supports are intended to be true scaffolds, in that students' reliance upon them fades over the course of instruction as they gain additional experience capturing and interpreting thermal camera images and videos. In the course of a coherent instructional unit, students then repeatedly draw upon this experience in order to design and conduct investigations of phenomena.

5.4 Student Learning About Energy Using Thermal Cameras

As an author team, we have conducted several studies investigating how thermal cameras support student learning about energy in the context of coherent project-based science units. In this section, we briefly summarize these studies.

5.4.1 Thermal Cameras in an Out-of-School Science Program

In the context of a science-focused afterschool program for middle school students, Nordine & Weßnigk engaged students in a series of connected activities using thermal cameras that took place in four two-hour sessions spread over the course of several months. The sessions were adapted from the energy unit in the "Investigating and Questioning Our World through Science and Technology" curriculum (Fortus et al., 2013). This project-based unit asks the driving question "Why do some

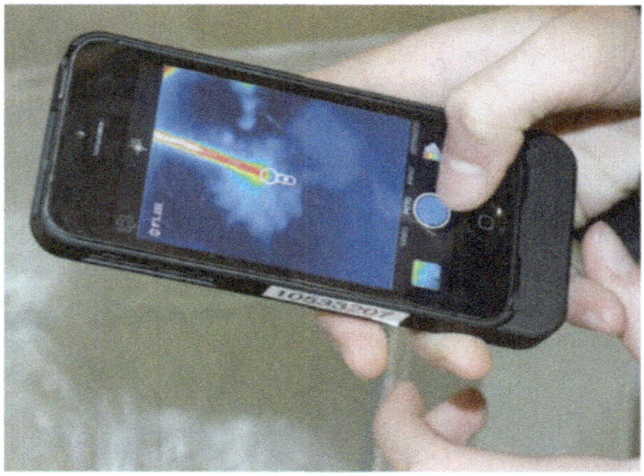

Fig. 5.2 Thermal imaging of hot water poured into an aquarium of room-temperature water

things stop and others keep going?", and emphasizes the central role of energy transformations in making sense of non-idealized, everyday phenomena (e.g., a plucked guitar string or a spinning top). Students learn to identify various energy forms based on the existence of "indicators" for each form (e.g., motion for kinetic energy) and to track whether these forms increase or decrease during phenomena. By investigating a series of everyday phenomena, students gather evidence that any time one energy form seems to increase, at least one other decreases (and vice versa). Empirical research has demonstrated that this approach effectively supports middle school students in developing ideas about energy transformation and transfer, but that students continue to struggle with ideas related to energy conservation (Fortus et al., 2015; Nordine et al., 2011). The afterschool program used the same driving question and included the same emphasis on tracking energy transformations in non-idealized, everyday phenomena.

In the first session, students explore a range of devices and energy toys (e.g., a candle-powered seesaw) and use their existing knowledge about energy to discuss how and why these devices/toys operate as they do. After this, students begin to explore phenomena that involve kinetic energy and gravitational energy, including pendula, falling objects, and student-designed marble rollercoasters. Through such investigations, students begin to develop the principle that any time one form of energy seems to decrease, at least one other must increase. At this point, students are ready to be puzzled by phenomena in which energy seems to be missing. Importantly, they have not yet used thermal cameras in their explorations.

To begin exploring phenomena in which energy seems to be missing, they first observe about 200 mL of hot water poured into a large aquarium (>10 L) of room-temperature water. Using a traditional thermometer and their own hands, they notice that the temperature of the aquarium does not measurably change. They then observe this same phenomenon through a thermal camera and begin to notice transient areas of local heating as the hot water is being poured (see Fig. 5.2).

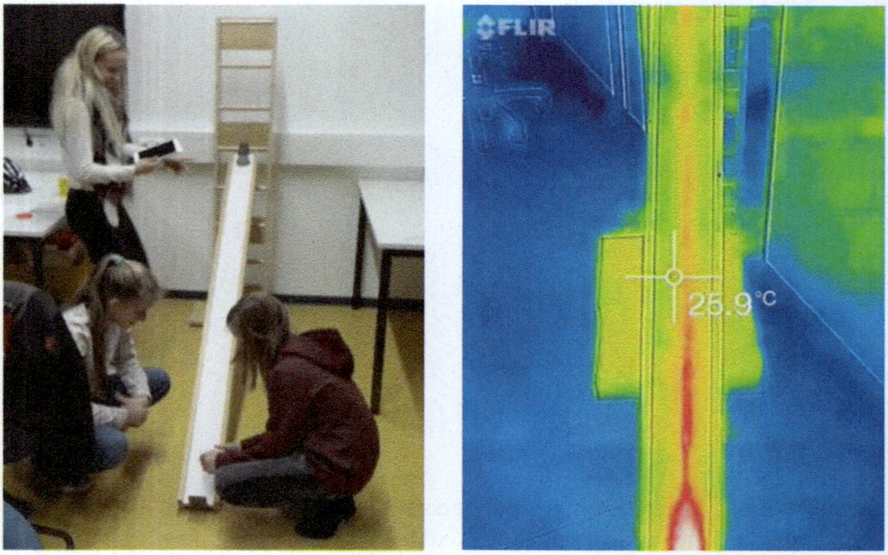

Fig. 5.3 Student groups using a thermal camera to view a cart sliding down a ramp (left), and a screenshot of a thermal imaging video showing "thermal tracks" as the sliding cart moves down the ramp

Students also use thermal cameras to investigate water being poured into a small (about 250 mL) and medium (about 500 mL) size container, this time observing with the thermal cameras that the temperature of the container rises. Through these investigations, students gather direct evidence of energy spreading and observe that the larger the system to which energy is transferred, the smaller the resulting change in temperature of that system. These observations set the stage for a subsequent investigation involving "missing" energy.

After using thermal cameras to directly observe energy dissipation as hot water is poured into a reservoir, students consider why a cart rolls down a ramp faster than it slides down the ramp. They first observe the phenomenon and record their initial ideas about why – if the cart begins with the same gravitational energy in each case – the cart has more kinetic energy at the bottom of the ramp when rolling than when sliding. Students are then provided with materials to investigate this question, included in the student materials is a thermal camera. As students explore the phenomenon, they inevitably view the rolling and sliding cart through the thermal camera (see Fig. 5.3).

The thermal camera provides students with compelling evidence that the "missing" energy when the cart slides down the ramp was in fact transferred to the track and quickly dissipated into the surroundings; students draw upon their recent investigation into the "missing" energy when hot water is poured into an aquarium to recognize that the track quickly cools as energy is transferred to the surrounding air.

As a culminating activity, students construct a Rube Goldberg Machine, which is an overly complex machine designed to perform a relatively simple task, like pushing a button. Students design and construct their own machine and explain the energy transfers and transformations as it operates.

To investigate how well this program supported student learning about energy conservation and degradation, Weßnigk & Nordine (2017) conducted semi-structured interviews before and after instruction and collected student feedback about their thermal camera use through surveys. In surveys, students reported being interested in the thermal cameras and reported learning key ideas about energy dissipation and conservation. For example, one student reported learning, "That heat, which I thought, doesn't disappear, but rather spreads out until everything is at about the same temperature". In semi-structured interviews, students were asked to use energy ideas to explain various phenomena, and students' responses were coded according to the complexity of understanding demonstrated according to four energy aspects (Duit, 1986): energy forms, transformation, degradation/dissipation, and conservation. Interviews showed a clear increase in student understanding of energy degradation/dissipation.

Thermal cameras are particularly useful for exposing the small, short-lived temperature changes that dissipate energy to the surroundings and obscure energy conservation in macroscopic phenomena. By using thermal cameras in the course of a set of coherent instructional activities, students demonstrated an increased understanding of energy degradation/dissipation as they used energy ideas to explain everyday phenomena; in reflecting upon the role of thermal cameras in their learning, students reported that the thermal cameras provided evidence that energy doesn't disappear, but rather "spreads out" in the environment.

In the teaching/learning sequence used Weßnigk & Nordine (2017), a coherent set of learning activities primes students to be puzzled by phenomena that seem to be missing energy, and thermal cameras play a key role in providing evidence that energy is not gone, but rather transferred to the surroundings.

5.4.2 Exploring the Role of Media Supports in Enhancing Student Learning Using Thermal Cameras

Building on the experiments from IQWST, Greinert and Weßnigk (2019) developed an energy course for the initial lessons in seventh grade (fourteen 90-minute lessons), which focuses on energy transfer to the environment during conversion and dissipation processes and is supported by IR cameras. In each of these processes, a part of the energy is transmitted into the environment and transported further from there. This focus enables learners to both recognize qualitative evidence for conservation and associated dissipation (Millar, 2014). In addition, this approach ties in with the concept of energy transfer, which is more understandable for learners (Behle & Wilhelm, 2017; K. Neumann et al., 2010).

However, the energy transfer to the environment during dissipative processes or the associated temperature increase is difficult to observe with traditional measuring instruments. Precisely this missing indicator significantly contributes to the difficulties of understanding energy dissipation processes (Daane et al., 2015). Making the transfer or the temperature increase visible can, however, help conceptualize thermal processes (Duit, 2014; Erickson, 1979; Millar, 2014; see also Yeo & Zadnik, 2001). We have therefore decided to use thermal cameras here as well.

The aim of the project "Making Energy Dissipation Tangible" is to support the students in developing their understanding of energy dissipation by focusing on the transfer of energy into the environment during dissipation processes and by using the thermal camera to visualize the processes.

The use of thermal cameras in teaching-learning processes has been met with high expectations, for example, Kleefeld & Bohrmann-Linde (2019, p. 209) argued that the presentation of measurement results as two-dimensional graphics might make processes such as heat conduction more accessible to students, allowing them to grasp and interpret measured values more easily, thereby reducing the risk of misconceptions. However, the real learning benefit remains subject to review.

An investigation of the learning benefits of using thermal cameras in an energy instructional unit took place within the framework of the studies EmIR (Energieentwertung mit Infrarot [Energy dissipation with Infrared]) and EmIR+ (the "+" sign indicates the inclusion of additional media didactic supports for using thermal cameras). A total of four groups were compared in an approximate 2x3 design: Group A was traditionally taught according to the core curriculum, Group B was taught according to the newly developed energy course and Group C was taught according to the newly developed energy unit using thermal imaging cameras. Group D had the same basic conditions as Group C and also received a multimedia didactic introduction to the interpretation and recording of thermal images and thermal videos (Greinert, 2020). Figure 7.4 shows the understanding of energy dissipation or energy conservation at posttest time, after working with the thermal imaging cameras. The mean value at pretest time, before the cameras were used, was marked as a red line. It corresponds to the mean value of the groups A, B and D tested at this time. Group C has the same socio-economical background as the other three groups, so that the mean value was also used as a comparative value.

The groups in the course developed a significantly better understanding of energy dissipation compared to the traditional teaching Group A (see Fig. 5.4). It should be pointed out that Group D, which worked in the course with media-didactic introduction to thermal imaging cameras, showed a significantly better understanding of energy conservation than the course group without IR camera as well as the traditional energy teaching (Greinert, 2020). This result suggests that the targeted use of IR cameras can support the development of an understanding of energy conservation.

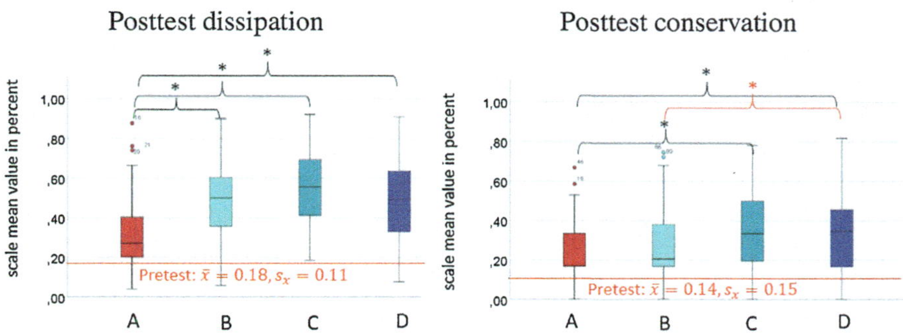

Fig. 5.4 Student learning about energy dissipation and conservation in four treatment conditions: traditional instruction (Group A), treatment instruction without thermal cameras (Group B), treatment instruction with thermal cameras (Group C), and treatment instruction with thermal cameras and embedded media-didactic supports (Group D)

All our previous studies have shown a great level of student interest in working with thermal imaging cameras. In this context, Weßnigk and Haase (2020) succeeded in showing that working with thermal imaging cameras particularly addresses the emotional component of situational interest. In addition, we found that seeing something with thermal imaging cameras, which is actually invisible, serves to capture students' interest. While the connection between thermal camera usage and student interest is influenced by a host of other factors, there is encouraging evidence that using thermal cameras can draw students into a lesson and engage them in using authentic tools to explore meaningful problems. Loukomies, et al., (Chap. 6) further discuss the potential of thermal cameras to support student interest and learning-related emotions.

5.5 Summary and Next Steps

Thermal cameras can play a key role in supporting students in learning about energy degradation/dissipation and conservation by providing direct evidence of small, short-lived energy transfers that occur as a part of most macroscopic phenomena. Research into the students' ideas about energy has shown that students' everyday experiences often do not align with normative ideas about energy degradation/dissipation and conservation, and that most students struggle to understand these ideas even by the end of secondary school.

To realize the potential of thermal cameras to support student learning about energy, they should be carefully used in the context of coherent science instruction. Sequences of learning activities should generate a need-to-know about energy degradation/dissipation/conservation when students have the prior knowledge to learn about them, at which point students are well-positioned to use evidence gathered via thermal cameras to collaboratively construct these ideas. In addition to strategically embedding thermal cameras in coherent instruction, instruction should include explicit scaffolds for students to independently and properly use thermal cameras to gather and interpret data, as well as to learn to interpret and use thermal camera evidence to construct arguments relating to energy conservation.

Through empirical investigations, we have found that when students are supported in using thermal cameras to learn about energy in the context of coherent instruction, even middle school students can begin constructing and using ideas about energy degradation/dissipation and conservation to make sense of phenomena.

The principle of energy conservation is among the most powerful ideas in science and engineering, bridging across science disciplines as well as socio-scientific issues that students encounter outside of school. Thus far, our empirical work has focused on the role of thermal cameras to explore physical phenomena that do not involve chemical systems, living systems, or engineering design, yet thermal imaging shows promise in helping students use energy to make sense of these domains as well (Kubsch et al., 2017; Wong & Subramaniam, 2020; Xie et al., 2010; Xu et al., 2019). Given the importance of energy across the sciences and the promise of thermal imaging to provide direct evidence of previously-hidden energetic processes, science education researchers and practitioners could benefit from further investigation of how thermal imaging may support students in building a coherent set of energy ideas that span disciplinary boundaries and set the stage for students to more successfully use energy to make sense of systems and phenomena that they encounter in their lives outside of school.

References

Alonzo, A. C., & Gotwals, A. W. (Eds.). (2012). *Learning progressions in science: Current challenges and future directions*. Sense Publ.

Behle, J., & Wilhelm, T. (2017). Aktuelle Schülerrahmenkonzepte zur Energie. *PhyDid B – Didaktik der Physik: Beiträge zur DPG-Frühjahrstagung*, 99–107.

Bransford, J. D., & Schwartz, D. L. (1999). Rethinking transfer: A simple proposal with multiple implications. *Review of Research in Education, 24*, 61–100.

Chen, R. F., Eisenkraft, A., Fortus, D., Krajcik, J., Neumann, K., Nordine, J., & Scheff, A. (Eds.). (2014). *Teaching and learning of energy in K – 12 education*. Springer. http://link.springer.com/10.1007/978-3-319-05017-1.

Daane, A. R., McKagan, S. B., Vokos, S., & Scherr, R. E. (2015). Energy conservation in dissipative processes: Teacher expectations and strategies associated with imperceptible thermal energy. *Physical Review Special Topics - Physics Education Research, 11*(1) https://doi.org/10.1103/PhysRevSTPER.11.010109.

de Lange, F. P., Heilbron, M., & Kok, P. (2018). How do expectations shape perception? *Trends in Cognitive Sciences, 22*(9), 764–779. https://doi.org/10.1016/j.tics.2018.06.002.

Driver, R., Squires, A., Rushworth, P., & Wood-Robnson, V. (1994). *Making sense of secondary science: Research into children's ideas*. Routledge.

Duit, R. (1986). *Der Energiebegriff im Physikunterricht [The energy concept in physics instruction]* (Institute for Science Education Series). IPN - Leibniz Institute for Science and Mathematics Education.

Duit, R. (2014). Teaching and learning the physics energy concept. In R. F. Chen, A. Eisenkraft, D. Fortus, J. Krajcik, K. Neumann, J. Nordine, & A. Scheff (Eds.), *Teaching and learning of energy in K – 12 education* (pp. 67–85). Springer. https://doi.org/10.1007/978-3-319-05017-1_5.

Duit, R., & Treagust, D. F. (2003). Conceptual change: A powerful framework for improving science teaching and learning. *International Journal of Science Education, 25*(6), 671–688. https://doi.org/10.1080/09500690305016.

Erickson, G. L. (1979). Children's conceptions of heat and temperature. *Science Education, 63*(2), 221–230. https://doi.org/10.1002/sce.3730630210.

Fortus, D., & Krajcik, J. (2012). Curriculum coherence and learning progressions. In B. J. Fraser, K. Tobin, & C. J. McRobbie (Eds.), *Second international handbook of science education* (pp. 783–798). Springer. http://www.springerlink.com/index/10.1007/978-1-4020-9041-7.

Fortus, D., Abdel-Kareem, H., Chen, J., Forsyth, B., Grueber, D., Nordine, J., & Weizman, A. (2013). *Why do some things stop while others keep going?* Activate Learning.

Fortus, D., Sutherland Adams, L. M., Krajcik, J., & Reiser, B. (2015). Assessing the role of curriculum coherence in student learning about energy. *Journal of Research in Science Teaching, 52*(10), 1408–1425. https://doi.org/10.1002/tea.21261.

Greinert, L. (2020). *Energieentwertung erfassbar machen: Konzeption und Evaluation eines curriculumorientierten Lehrgangs mit Fokus auf Energietransfer und Visualisierung mittels Infrarotkamera*. Universität Hannover.

Greinert, L., & Weßnigk, S. (2019). Energieentwertung mit der IR-Kamera – Studie zum Einfluss der IR-Kamera auf das Energieverständnis in einem curriculumorientierten Lehrgang mit Fokus auf Energietransfer. *Zeitschrift für Didaktik der Naturwissenschaften, 25*(1), 245–257. https://doi.org/10.1007/s40573-019-00102-w.

Greinert, L., & Weßnigk, S. (2020). Unsichtbares sichtbar machen: Interpretation von Wärmebildern mit Hilfe eines WBK-Führerscheins. *Digital Unterrichten Biologie, 1*, 8–9.

Haglund, J., Jeppsson, F., Hedberg, D., & Schönborn, K. J. (2015). Thermal cameras in school laboratory activities. *Physics Education, 50*(4), 424–430. https://doi.org/10.1088/0031-9120/50/4/424.

Herrmann-Abell, C. F., & DeBoer, G. E. (2018). Investigating a learning progression for energy ideas from upper elementary through high school. *Journal of Research in Science Teaching, 55*(1), 68–93. https://doi.org/10.1002/tea.21411.

Hoppe, T. (2018). *Interpretation von Erkenntnisgewinnung aus Wärmebildern*. Universität Hannover.

Kali, Y., Linn, M. C., & Roseman, J. E. (Eds.). (2008). *Designing coherent science education: Implications for curriculum, instruction, and policy*. Teachers College Columbia University.

Kleefeld, S., & Bohrmann-Linde, C. (2019). Die Wärmebildkamera im naturwissenschaftlichen Unterricht. *MNU Journal, 72*(3), 209–216.

Köhn, K. (2019). *Der IR–Experimentierführerschein als mediendidaktische Einbettung für den Physikunterricht in der Mittelstufe*. Universität Hannover.

Krajcik, J. S., & Czerniak, C. L. (2018). *Teaching science in elementary and middle school: A project-based learning approach* (5th ed.). Taylor & Francis Group: Routledge.

Kubsch, M., Nordine, J., & Hadinek, D. (2017). Using smartphone thermal cameras to engage students' misconceptions about energy. *The Physics Teacher, 55*(8), 504–505. https://doi.org/10.1119/1.5008354.

Lederman, N. (2007). Nature of science: Past, present, and future. In S. K. Abell & N. Lederman (Eds.), *Handbook of research on science education* (Vol. 2, pp. 831–879). Lawrence Erlbaum Associates.

Liu, X., & McKeough, A. (2005). Developmental growth in students' concept of energy: Analysis of selected items from the TIMSS database. *Journal of Research in Science Teaching, 42*(5), 493–517.

Millar, R. (2014). Towards a research-informed teaching sequence for energy. In R. F. Chen, A. Eisenkraft, D. Fortus, J. Krajcik, K. Neumann, J. Nordine, & A. Scheff (Eds.), *Teaching and learning of energy in K – 12 education* (pp. 187–206). Springer International Publishing. https://doi.org/10.1007/978-3-319-05017-1_11.

National Academies of Sciences, Engineering, and Medicine. (2018). *How people learn II: Learners, contexts, and cultures*. National Academies Press. https://doi.org/10.17226/24783.

National Research Council. (2000). *How people learn: Brain, mind, experience, and school: Expanded edition*. National Academies Press. https://doi.org/10.17226/9853.

Netzell, E., Jeppsson, F., Haglund, J., & Schönborn, K. J. (2017). Visualising energy transformations in electric circuits with infrared cameras. *School Science Review, 98*(364), 19–22.

Neumann, S., & Hopf, M. (2012). Students' conceptions about 'radiation': Results from an explorative interview study of 9th grade students. *Journal of Science Education and Technology, 21*(6), 826–834. https://doi.org/10.1007/s10956-012-9369-9.

Neumann, K., Viering, T., & Fischer, H. (2010). Die Entwicklung physikalischer Kompetenz am Beispiel des Energiekonzepts. *Zeitschrift Für Didaktik Der Naturwissenschaften, 16*, 285–298.

Neumann, K., Viering, T., Boone, W. J., & Fischer, H. E. (2013). Towards a learning progression of energy. *Journal of Research in Science Teaching, 50*(2), 162–188. https://doi.org/10.1002/tea.21061.

Nordine, J. (Ed.). (2016). *Teaching energy across the sciences, K-12*. National Science Teachers Association.

Nordine, J., Krajcik, J., & Fortus, D. (2011). Transforming energy instruction in middle school to support integrated understanding and future learning. *Science Education, 95*(4), 670–699. https://doi.org/10.1002/sce.20423.

Nordine, J., Krajcik, J., Fortus, D., & Neumann, K. (2019). Using storylines to support three-dimensional learning in project-based science. *Science Scope, 42*(6), 85–91.

Schneider, B. L., Krajcik, J. S., Lavonen, J., & Salmela-Aro, K. (2020). *Learning science: The value of crafting engagement in science environments*. Yale University Press.

Smith, J. P., DiSessa, A., & Roschelle, J. (1993). Misconceptions reconceived: A constructivist analysis of knowledge in pieces. *Journal of the Learning Sciences, 3*(2), 115–163.

Solomon, J. (1983). Messy, contradictory, and obstinately persistent: A study of children's out-of-school ideas about energy, *65*(231), 225–229.

Trumper, R. (1993). Children's energy concepts: A cross-age study. *International Journal of Science Education, 15*(2), 139–148.

Watts, D. M. (1983). A study of schoolchildren's alternative frameworks of the concept of force. *European Journal of Science Education, 5*(2), 217–230. https://doi.org/10.1080/0140528830050209.

Weßnigk, S., & Haase, J. S. (2020). Thermal imaging cameras as "catch factors" for students' current interest. *Progress in Science Education, 3*(1), 36–53. https://doi.org/10.25321/prise.2020.966.

Weßnigk, S., & Nordine, J. (2017). Die Bedeutung von Unterrichtslehrgängen für die Entwicklung von Kompetenz: Empirisher Beleg für die Wichtigkeit von Lehrgangsentwicklung sowie ein Beispiel für einen Lehrgang zur Energieentwertung. *Plus Lucis, 2*, 9–14.

Wong, C. P., & Subramaniam, R. (2020). Use of technology in biology education – Case of infrared thermal imaging. *Journal of Biological Education*, 1–13. https://doi.org/10.1080/00219266.2020.1808510.

Xie, C., Hazzard, E., & Nourian, S. (2010). Infusing technology into engineering education. *Engineering and Design Education Research Summit, P-12*, 1–14.

Xu, X., Wu, M., & Wang, X. (2019). Smartphone visualization of thermal phenomena with thermal imaging accessories. *Journal of Chemical Education, 96*(11), 2545–2552. https://doi.org/10.1021/acs.jchemed.9b00131.

Yeo, S., & Zadnik, M. (2001). Introductory thermal concept evaluation: Assessing students' understanding. *The Physics Teacher, 39*(8), 496–504. https://doi.org/10.1119/1.1424603.

Chapter 6
Students' Emotions Related to Thermal Camera Activities in Primary Science Lessons

Anni Loukomies, Taina Makkonen, Jari Lavonen, and Kalle Juuti

6.1 Introduction

In this research project, we examine the emotions reported by students when they engage in thermal camera activities. We also examine how students experience learning situations in which thermal cameras are employed. Digital devices such as thermal cameras allow students to observe and study abstract phenomena that are difficult to grasp with traditional instructional methods. Using such tools may also support positive emotional aspects of learning, such as situational interest, that is awakened by an external factor that captures the attention, and precedes personal interest in interest development process (Hidi & Renninger, 2006).

6.1.1 Emotions in Learning Situations

Students' performance is influenced by emotions, motivation, learning strategies, and competence beliefs (Pekrun et al., 2010; Putwain et al., 2018). Task-related positive emotions may support students in reaching the objectives of learning tasks, whereas negative emotions may generate obstacles to learning (Pekrun et al., 2018). Teachers are responsible for inspiring excitement about learning in the context of the discipline, in addition to familiarising students with provided content (Pekrun et al., 2018). A teacher's choice of lesson activities and his or her interactions with the

A. Loukomies (✉) · J. Lavonen · K. Juuti
Faculty of Educational Sciences, University of Helsinki, Helsinki, Finland
e-mail: anni.loukomies@helsinki.fi; jari.lavonen@helsinki.fi; kalle.juuti@helsinki.fi

T. Makkonen
Viikki Teacher Training school, Faculty of Education, University of Helsinki, Helsinki, Finland
e-mail: taina.makkonen@helsinki.fi

J. Haglund et al. (eds.), *Thermal Cameras in Science Education*, Innovations in Science Education and Technology 26, https://doi.org/10.1007/978-3-030-85288-7_6

students may support students' positive emotions towards a topic or subject. Support for positive learning related emotions is especially important in the context of school science, as students generally find science to be interesting, but most students do not find science and technology careers interesting (Shirazi, 2017).

According to the Finnish National Core Curriculum for Basic Education (FNC-CBE, 2014), "the interests, appraisals, working approaches and emotions of the pupils, as well as their experiences and ideas of themselves as learners, influence their learning process and motivation" (p. 17). Emotions encompass affective, cognitive, motivational, physiological, and expressive components, of which the affective component is always present, whereas the other components may not exist (Pekrun et al., 2018).

Students experience a wide range of emotions in their academic studies (Pekrun et al., 2002). They are referred to as academic emotions (Pekrun et al., 2018). Academic emotions can be classified positive/negative, or activating/deactivating (Pekrun et al., 2002, p. 97). Further, they can be categorized into subcategories of achievement emotions, epistemic emotions, topic emotions, and social emotions. Achievement emotions can be approached from activity or outcome perspectives. Boredom, for example, is an achievement emotion related to the activity at hand. Epistemic emotions are related to knowledge or knowledge generation, and they express the affective side of cognitive processes. In the knowledge generation process, situational interest follows curiosity, and it can be referred to as an epistemic emotion (Pekrun et al., 2018). Topic emotions, in turn, are aroused by the learning material or topic at hand, but they are not necessarily related to the learning process. Furthermore, social relationships are usually involved in learning situations, and they generate many kinds of emotions. These emotions are referred to as social emotions, and they may not always be related to achievements but rather to the relationships of the persons in the learning situation (Pekrun et al., 2018). A science teacher is one "knot" in the social network of the class and may have an influence on students' emotional experiences.

Interest can be included among the emotions because it encompasses all the typical components that emotions have, namely physiological changes, facial and vocal expressions, patterns of cognitive appraisal, a subjective feeling, and an adaptive role across the lifespan (Silvia, 2008). Boredom can be regarded as the opposite emotion to interest. It relates to reduced arousal, lack of motivation, attention problems, and impaired performance (Pekrun et al., 2010). Interest may be accompanied by other positive or negative emotions (Ainley & Hidi, 2014), depending on how and why it emerges. Therefore, the expression of interest may vary.

Interest is associated with cognitive and attention-related aspects as well (Lonka & Ketonen, 2012). In the person-object theory of interest, interest is not defined as an emotion but as 'a unique motivational variable, as well as a psychological state that occurs during interactions between persons and their objects of interest, and [it] is characterised by increased attention, concentration and affect' (Hidi, 2006, p. 70). External factors that capture attention may awaken situational interest (Schiefele, 1991). Hidi and Renninger (2006) suggest that aroused (caught) and maintained (held) situational interest first develop into emerging personal interest and then into well-developed personal interest. Whether or not interest is included in the emotions, there is a consensus that interest promotes learning, exploration and information seeking (Silvia, 2008; Ainley & Hidi, 2014) and therefore, it is an important means for a science teacher.

6.1.2 Using Thermal Cameras in the Primary Classroom

Thermal cameras or infrared (IR) cameras have shown considerable potential in promoting science learning across multiple levels of education (Haglund et al., 2016a, 2016b). We argue that the potential of thermal cameras to illuminate aspects of life that are otherwise invisible and show novel aspects of a phenomenon may spark the curiosity and situational interest of students (Palmer et al., 2016). In recent years, the cost of thermal cameras has decreased substantially, allowing the integration of this user-friendly, non-contact real-time visualisation technology in ordinary schools. With these instruments, students can visually observe otherwise invisible thermal phenomena, supporting the learning of concepts typically considered difficult in traditional instruction (Haglund et al., 2015). In addition to complementing the measurements traditionally made with thermometers, IR cameras provide students with an instant and dynamic approach to investigating various scientific phenomena (Xie, 2011), for example, the opportunity to measure small temperature changes without touching an object.

Thermal cameras create interesting opportunities for primary-level science education. First, they offer possibility to build on students' everyday experiences and simple classroom observations (Haglund et al., 2016a, 2016b). Second, they allow for the multisensory exploration of thermal phenomena, combining, for example, visual, tactile, and bodily-kinesthetic experiences (Larsson et al., 2019). Third, IR cameras are helpful in producing meaningful cognitive conflicts (Schönborn et al., 2014), and fourth, they encourage students to discuss science and engage in true dialogue (Dolo et al., 2018). Furthermore, given that temperatures are indicated with an intuitive colour scale on the camera screen, even very young students can easily access the visual information.

A wide range of applications have been suggested for the pedagogical use of thermal cameras, including classroom experiments in thermodynamics (e.g., Xie & Hazzard, 2011), mechanics and electromagnetism (Vollmer & Möllmann, 2010), and chemistry (Short, 2012; Xie, 2011). Many of these activities are targeted at secondary and university students, but some may well be applied at lower levels. For example, the Next Generation Science Standards (NGSS, 2013) emphasise that students at lower levels must learn to make observation on heat transfer. The Finnish National Core Curriculum for Basic Education mentions temperature measurements, as well as learning the basics of thermal energy and energy conversions as ways to introduce grade 3–6 students to the law of conservation of energy (FNCCBE, 2014).

6.1.3 Instant Video Blogging as an Experience Sampling Method

Teachers require accurate information about students' emotions to support their learning and motivation and adjust their teaching actions. According to Shirazi (2017), in addition to concentrating on teaching conceptual knowledge, science teachers should also consider the way in which subject content knowledge is taught. Having a teacher with a good sense of humour and empathy increases positive experiences in learning conceptual knowledge. If students can express their affects, even negative ones, and the teacher takes such student perspectives into account, students' autonomy is supported (Reeve & Halusic, 2009). As a teacher interacts with students in a science class, it is plausible that his or her perception of the students' emotional states can influence his or her decisions throughout the lesson. A teacher may obtain hints regarding students' emotional states by recognising their facial and verbal expressions, but it may be difficult for the teacher to draw the correct conclusions through observations. For example, Loukomies et al. (2015) found that students may hide their actual emotional states from the teacher. Therefore, the teacher may draw incorrect conclusions about the students' level of interest. Even with familiar students, a teacher may experience difficulty interpreting their emotional states (Loukomies et al., 2015). It is essential to obtain accurate information about what students actually experience to design science teaching sequences that could promote positive emotions and engagement. Finding avenues for collecting students' experiences in the situation is critical because if students are retrospectively asked about situational experiences, they may interpret the past incorrectly due to memory bias (Shirazi, 2017). To overcome these challenges, we applied an experience sampling method (ESM) in the data collection reported in this chapter. Traditionally, ESM is defined as repeatedly collecting written responses to questions tailored to fit the particular situation (Hektner et al., 2007). By employing ESM, it is possible to separate longer-term conditions from the immediate context of a certain feeling (Csikszentmihalyi & Hunter, 2003). Usually, experiences are reported over a period of several days, multiple times per day (Katz-Buonincontro & Hektner, 2014). After hearing a signal, the participants evaluate their experiences at that moment (Csikszentmihalyi & Hunter, 2003). Traditionally, ESM studies employ paper-and-pencil questionnaires, which students complete during the lesson after specific activities (in a science education context, see Palmer, 2009); also digital tools have been designed for mobile phones or iPads as a means of reporting experiences (Litmanen et al., 2012; Katz-Buonincontro & Hektner, 2014). We have developed an Instant Video Blogging (IVB) procedure as a digital ESM method to support the effortless collection of students' experiences in primary science classes. With IVB we mean that during the activity, students do video blogging that is structured with guiding questions and themes around which they speak. Students have a possibility to choose a quiet place in which they want to do the video blogging. The activity is not controlled, which means that the students also have possibility to refuse to speak.

6.1.4 Research Questions

In this study, we used thermal cameras in the context of primary science class. We introduce instant video blogging (IVB) as a method for collecting students' first-hand reports about their momentary emotions. Video blogging took place in science learning situations where students used thermal cameras and investigated phenomena related to temperature changes. In this research project, we examine the emotions reported by students when they engage in thermal camera activities. We also examine how students experience learning situations in which thermal cameras are employed. Our research question is as follows:

> What emotions did students report in science lessons related to thermal phenomena and thermal camera activities?

6.2 Methods

6.2.1 Participants and Procedure

Twenty-four Finnish fifth-grade students (11–12 years) took part in this study. The school in which this study took place is located in a suburban area of Helsinki, Finland. The study consisted of four 45-minute science lessons. For the purposes of IVB the students were given smartphones equipped with video cameras. The data were pseudonymised. The students were numbered according to the numbers of the phones they used. The phones were given to the students randomly. During the lessons, there were three adults present in the classroom. One of them was the teacher of the class, who also works as a researcher. The other two were researchers, who also joined in the organisation of the classroom activities and participated as co-teachers. All three adults interacted with the students and guided them during the lessons in thermal camera use, as well as in planning the investigations and communicating the observations. The teacher of the class had overall responsibility for the activities, and she ensured that the activities were in line with the requirements of the curriculum. Permission was obtained from the school principal and the guardians of the students to conduct the described activities.

In this study, we applied IVB as an experience sampling method. Students were asked to introduce their activities and evaluate their feelings based on the phases of the lesson. The sampling frequency was not randomised but followed the structure of the lesson. Specifically, the students were asked to explain how they felt in the situation. At the beginning of each IVB session, the students explained what they were doing.

6.2.2 Description of the Teaching Sequence

Lesson 1 began with students familiarising themselves with the smartphones. The teacher introduced the aims of the lesson (learn about heat transfer and insulation). Students were instructed to work in groups and make thermal camera observations of various objects. Students were instructed to discuss their observations with their group mates. From a group work perspective, students were to allocate tasks within the group evenly and allow everyone to use the tablet connected to the thermal camera. An overall aim was to learn about heat transfer and insulation. Another aim that was related to the videoblogging activity was to practice to communicate opinions and describe emotions. They were reminded of the correct use of the smartphone cameras. The students were also reminded that their video blogs would only be used for research purposes, not for academic evaluation or any other uses. The students were told that they should stop when advised, take the smartphone out, and answer the questions on video. Then the students were introduced to the categorization of academic emotions by Pekrun et al. (2010) (Fig. 6.1) and the concept of interest. This was done in order to provide students with emotion-related words they can use in their video-blogs. The first videoblogging was a rehearsal. It involved recalling a site visit to a university physics laboratory that took place during the previous school year and included thermal camera activities. The question that students answered was "What do you remember about the site visit to the laboratory in the department of physics?"

Lesson 2 began with signing into an interactive white board. The students were asked to think about emotions and send names of different emotions. Through discussion, the concept of emotion was defined. The discussion was an enrichment of the discussion about academic emotions in the previous lesson, and the topic was taken into students' world. It was concluded that emotions are reactions to something, but the brain can also generate an emotional state without a stimulus. The emotions mentioned by the students were categorised into the template of dimensions of academic emotions that is introduced in Fig. 6.1. After discussions about emotions, the student groups were given thermal cameras, one camera per four students. After obtaining the cameras, the students were asked to video-blog (recording 2) about the following questions: (1) What are you doing at the moment,

Fig. 6.1 Categorization of emotions by Pekrun et al. (2010)

Negative-positive	**Deactivating-activating**	
	Positive deactivating (e.g. relief, relaxation)	Positive activating (e.g. enjoyment, success)
	Negative deactivating (e.g. boredom, hopelessness)	Negative activating (e.g. anxiety, shame)

and (2) what are your emotions right now? The same questions were used in lessons 2–4, in videoblogs 2–5.

Lesson 3 began with instructions to walk around and check the temperatures of various objects. There was an electric water kettle in the class, as well as a cooling pack and frozen ice-cream. Students also observed one another with the infrared camera. After activities described above, the students video-blogged (recording 3) about the same questions as at the end lesson 2.

Next, another task was given to the students. They were told to measure the temperatures of certain objects and save their results using Microsoft Teams. Students were about to measure the temperatures of their hands and then do the same through plastic, window glass, aluminium foil, plastic wrapper, fabric, and other materials. After the measuring activity, the children were asked to reflect on the previous activity by videoblogging (recording 4) about the following questions: (1) What are you doing at the moment, and (2) what are your emotions right now?

Lesson 4 began with recalling the results of the previous lesson via a demonstration. The measurements were taken in a teacher-led fashion, with the task to determine whether a hand was visible on the thermal camera screen alone, as well as through plastic, foil, and glass. A demonstration about heat conduction followed. After the demonstrations, the students video-blogged about the same questions as were used in previous lessons (recording 5). Then, a teacher-led discussion followed regarding what windows are like in northern countries: they are triple-layered, and there is air between the glass panes for insulation purposes. The same phenomenon was discussed from the point of view of clothing. The pupils used their own jackets to check how the inner and outer surfaces of the garments felt. Then, a short introduction to various methods of heat transfer took place, namely that heat is transferred through thermal conduction, thermal radiation, and thermal convection. Observations were converted into a model of heat transfer processes. Examples were given regarding the various methods of heat transfer. Next, the students and teachers went to the schoolyard, with the task of examining clothing with the thermal camera. Some examples were given to inspire student activity, which included (1) with or without a hat, (2) with or without gloves, (3) coat zipped or unzipped, and (4) skin or skin after touching a metal surface. One additional task was to examine the temperature of various locations within the school building. The students chose their activities in a group, and at the end of the lesson, they video-blogged about the same questions as were used at the end of previous lessons (recording 6).

6.2.3 Data Collection and Analysis

The data for this study consist of transcribed recordings of the students' IVB videos. There were six video blogging situations, in which 24 students participated. Video recordings were transcribed verbatim, and one emotional expression in a recording was chosen as an analysis unit. In most recordings there were only one emotional expression, but in some of them there were 2–3 different mentions of emotions.

If one emotion is mentioned more than once in one recording, it is counted as one mention. Examples of mentioning the same emotion more than once in one recording, and mentioning different emotions in one recording are introduced below.

> *Well, we used thermal cameras again today and it was very* ***nice****, we examined the temperature of a tree, we did it outside and it was very* ***nice****.* (Student 5, Situation 6, one emotional mention)
>
> *In this lesson we did everything we've done before and filmed with thermal cameras. I don't know f***ing why we did the same, it was* ***boring****, I don't know why it is so* ***boring*** *but is it very* ***boooooriiiing*** *because I know so much about thermal cameras as I visited fire station. Bye.* (Student 1, Situation 4, one emotional mention)
>
> *Well, we looked if we see different things through the thermal camera and it was* ***nice****, it was quite* ***fun****, more fun than what we did in the previous lesson (Student 4, Situation 4, two emotional mentions)*

The transcriptions were read, keeping in mind the theoretical construct about emotions based on the research of Pekrun et al. (2018). All mentions were categorised, and therefore there are more analysis units than recordings. Inductive categories of different emotions were constructed. However, the research of Lonka and Ketonen (2012) shows that even among adults, the evaluations of different positive emotional aspects such as interest, enthusiasm, determination, and energy are all positively correlated with each other. Therefore, in this research, positive inductive categories were combined into main categories positive and negative. Besides these, there were neutral expressions that did not include emotional content. The video blogging situations and the number of successful recordings are presented in Table 6.1.

6.3 Results

Altogether, 144 potential measurements yielded 113 video blogs that 24 students recorded during the activities. Twenty-four video clips were recorded during the

Table 6.1 Video blogging situations and the number of successful recordings

Recording	Activity	Recordings
1, lesson 1	Rehearsal	24 (not included in the data)
2, lesson 2	Discussion about emotions and the ways of classifying them, familiarizing with the function of thermal cameras	21
3, lesson 3	Student-led group work with thermal cameras	20
4, lesson 4	Student-led group work with thermal cameras, looking at objects through medium	20
5, lesson 4	Teacher-led demonstration and discussion about looking at objects through medium.	16
6, lesson 4	Using thermal cameras outdoors	12

rehearsal round, and 89 were related to thermal camera activities and the emotions that emerged during those activities. These 89 video clips constitute the data, and one emotional mention was chosen as an analysis unit. The amount of successful videoblogs decreases along the teaching sequence. Some students may not have been present at all classes due to different reasons, but some of them just did not do the video-blogging when asked. The ethical research principles were followed carefully and the students were not pushed to do blogging, but they had the opportunity to refuse to speak. The distribution of the answers into main categories positive and negative and to their subcategories is presented in Table 6.2.

Positive verbal expressions dominated the answers, there were altogether 87 positive mentions. Positive expressions were most often related to the activity being 'nice' or 'fun' in the students' opinion. In fact, nice was mentioned 60 times, and fun 11 times. Typically, when students said something was nice or fun, they referred to a hands-on activity. Interesting was mentioned four times. Below two examples containing positive mentions are presented.

> *We used thermal cameras and tried everything... how warm...we were given a cold bag and ice cream, and then, we checked how warm they were, and then, we took a look at the water kettle, how warm it is, and I think this was nice, and I would like to do more similar tasks. It was nice.* (Student 13, Situation 3)
>
> *We did different tests with thermal cameras, some basic stuff, tried how warm different objects are, autonomous activities. I like it that I have to work by myself with not so much instruction. I think it was nice.* (Student 24, Situation 3)

There were 17 analysis units categorised negative. Negative verbal expressions were most often related to the student being bored without explanation and also the activity being boring because of technical problems and the thermal camera not working the way it was intended to. Some students were even bored because the teaching sequence took two days and they experienced it as boring to continue with the same topic. In addition to boredom, students also mentioned that the activity

Table 6.2 Frequencies of emotional mentions in the students' video blogs in 89 video clips

Main category	Positive emotional mentions	*Total 87*
Subcategory	*Nice*	*60*
	Good feeling	*4*
	Fun	*11*
	Happy	*1*
	Pleasant	*2*
	Glad	*3*
	Interesting	*4*
	Ok	*2*
Main category	**Negative emotional mentions**	***Total 17***
Subcategory	*Boring*	*12*
	Not nice	*2*
	Irritating	*2*
	Scary	*1*

was irritating or that they felt irritated. Below, there are examples of excerpts with negative mentions.

> *We worked with thermal cameras, and our group did not do well, because the thermal camera kind of got broken, and it was irritating, and the rest of the group did not go to where they were supposed to go.* (Student 23, Situation 3)
>
> *We spoke about [the] thermal camera activities of the previous lesson and tested them again, or the teachers showed it again. It was a repetition of yesterday, and it was quite boring to do the same thing again.* (Student 13, Situation 5)

The content of an analysis unit was considered neutral if there was no emotional content present. In these cases, the students only described the activity and skipped the request to describe their emotional state. In total, 12 out of 89 video clips encompassed only neutral content. For example, consider the following extract:

> *We were outside with the thermal cameras and tested everything, how warm the jacket is, and we noticed that if you don't have your cap on, your head is radiating much more heat, but it is kind of a bad thing because then, the heat goes away, and this is my observation.* (Student 13, Situation 6)

There were 12 situations in which a video blog contained both positive and negative mentions. Then it was categorised into both positive and negative categories. The most typical combination of positive and negative content was a combination of nice and boring. An example of such case is as follows.

> *Well now we use thermal cameras, and I don't know the idea of it but maybe one learns to use it. It is quite* ***nice*** *but we also did it yesterday so it is a bit* ***boring*** *but* ***nice*** (Student 4, Situation 5)

What the students said about what they were doing reveals that the student-led thermal camera activities cannot ensure correct conceptual understanding of the phenomenon. The following excerpt reveals misunderstandings in the student's knowledge structure that would require significant teacher-led discussions to correct. Figure 6.2 visualises the situation.

Below there is an example of a misconception related to the situation visualised in Fig. 6.2.

> *It is difficult for me to explain. We had thermal cameras, and then, we saw some kind of spirit through the glass, even though there was nobody, only a hand, and it was actually quite scary, but it was nice.* (Student 8, Situation 3)

In one video blog, the same situation was explained in a more correct way.

> *Well, it was nice, we examined if thermal cameras can go through things, and it was interesting that one cannot see through everything, for example glass, but the camera shows the mirror image of the one looking at the glass with the thermal camera.* (Student 14, Situation 3)

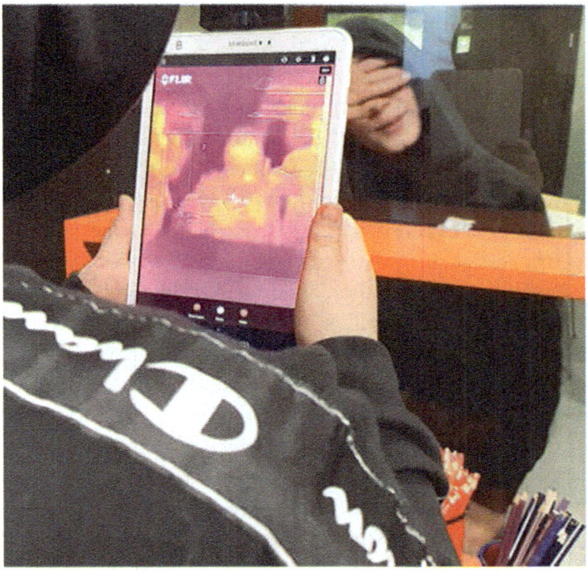

Fig. 6.2 Three students observe a student behind the glass. The glass reflects heat radiation coming from the observers, and students see thermal signatures of themselves on the tablet screen

6.4 Discussion

The main aim of this research was to examine what kinds of emotions students report having experienced during thermal camera activities. The positive emotions expressed in IVB were mostly related to the activities being 'nice'. In other words, the students' speech was not detailed; it seems that they referred to all positive emotional aspects of the situation as nice. The fact that their speech included both positive and negative content confirms that their ability to differentiate between different emotions is not very well developed, even though the topic was considered before the video blogging started, using the framework of Pekrun et al. (2018). This is in line with previous research in that even among adults, there is correlation between experienced positive emotions (Lonka & Ketonen, 2012).

Interest was mentioned three times in students' answers. The potential of thermal cameras to awaken interest is a topic that should be considered further. Support for situational interest is important because situational interest may develop and transform into personal interest if maintained (Hidi & Renninger, 2006). Interest-based interactions with the environment provide modes that combine positive cognitive and affective qualities (Krapp & Prenzel, 2011), which motivate people to learn (Silvia, 2008). Therefore, activities should be chosen to support processes of interest development. According to Silvia (2008), the evaluation of the 'novelty and complexity' and 'comprehensibility' of an event are crucial for investing interest in it. The participants in this study had already used thermal cameras once, so the comprehensibility requirement was met. The new context of the activities, as compared with the previous occasion, and the different tasks provided novelty and complexity.

The negative affect expressed in the IVB data was mostly related to students being bored or irritated. Pekrun et al. (2002) claim that the experience of boredom may be related to too little or too much demand with respect to students' ability beliefs. When a student feels incapable of keeping up with demands and following the instructions, boredom may offer a way to escape from this uncomfortable situation (Pekrun et al., 2002, p. 94). Science is experienced as a difficult subject (Osborne et al., 2003), and it may be easier to claim it is boring than challenge oneself to examine the concepts and their relationships. In addition to teaching the content, which includes examining students' pre-existing knowledge and altering the teaching accordingly, teachers should encourage students to scrutinise their mindsets in relation to their interest in the subject. Students may have a fixed perspective of interest, which may manifest itself in their unwillingness to provide opportunities for new topics and learn to become interested in them. A growth approach to interest leads students to express interest in new areas (O'Keefe et al., 2018). Classroom interactions should support a growth approach to interest, and this might be supported by choosing appealing technologies, such as thermal cameras. The skill of developing interest is a starting point for a successful learning process. Supporting students' interest is emphasised in the Finnish Core Curriculum for Basic Education (FNCCBE, 2014). However, expressing and accepting negative emotions is important from the point of view of motivation (Reeve & Halusic, 2009). Students only reveal their negative emotions in an atmosphere of trust, and therefore, the teacher should welcome these expressions. If revealed, negative emotions can be discussed and elaborated further. Through discussions, students will learn how to deal with their emotions, change their perspectives, and find new, more constructive ways of reacting. Studying science is not always considered a positive experience. However, dealing with negative emotions and converting them into constructive behaviours, as well as maintaining perseverance and finding constructive ways of dealing with negative emotions, are important conditions for learning and also a science teacher's concern. If the science teacher acquires accurate information about the students' emotions and thoughts when using equipment such as thermal cameras, the use such devices can improved.

The topic *Heat transfer and its properties* is conceptually challenging, as revealed in the students' descriptions of what they were doing. Students' views may be very different from the teacher's intentions. Rosebery et al. (2010) claim that this heterogeneity is fundamental to teaching, and they see classrooms as spaces in which entire systems of meaning or ways of seeing the world come into contact with one another in both planned and unplanned ways (p. 351). Our results are in line with what Schönborn et al. (2014) argued after having asked 12–13-year-old students to utilise thermal cameras to explain why metals feel colder than wood at room temperature. They claim that although the participants could observe the changes in temperature, they were not able to provide a satisfactory explanation of heat transfer ("flowing") from their bodies to the studied objects. Schönborn et al. (2014) concluded that students need at least a basic level of conceptual knowledge of heat transfer before being able to interpret thermal images in an appropriate way. In order to reach a common understanding, thermal camera activities should be enriched with teacher-led discussions. In this sequence, teacher-led activities were reduced in order to increase student-led activities. The students could tell by looking at the camera screen that blue corresponded to "cold" and red to "hot", but they did not get further than that with the device. However, students liked working with thermal cameras, and the device even sparked three students' situational interest. Also, the thermal camera was found to be a useful aid for observations because it made the invisible visible. However, the observations should be guided in such a way that the teacher considers the pre-existing knowledge of the students and guides the observations in the intended direction. In teacher talk, observations and interpretations should be separated. Teacher encouragement helps engage students in discussions about their observations. The teacher should also encourage students to use modes of communication other than speech alone to express their observations, such as drawings and charts.

Also, there should be demonstrations included in the teaching sequence regarding various methods of heat transfer other than radiation. Thus, students can have a clearer view of what the thermal camera is actually measuring and the observations could be transferred into a model. The coloured images demonstrate the measured surface temperature of objects. Tests with an obstacle between the object and the camera illustrate the properties of radiation and its similarities and differences to light. This kind of an experiment is an example of a classification task.

However, according to Rosebery et al. (2010), symbolic and real-world referents were sometimes completely blended in the talk of the children. A dialogic approach might be useful in steering the students' view toward a scientifically correct one, and it even has the potential to support students' interest in science learning (Juuti et al., 2019). In their study of 9–11-year-olds, Haglund et al. (2016a) included a teacher-led introduction to heat flow in the teaching sequence on basic thermal phenomena, and they reported that the students were able to interpret the IR images and understand the heat-flow model when exploring heat conduction. Moreover, thermal imaging induced these students to spontaneously ask relevant what-if questions and answer them via experimentation. Likewise, Larsson et al. (2019) found that when engaging in open-ended heat-related activities that require planning on the part of students, the use of thermal cameras enabled fourth grade students to assess and adjust their experimentation. Given the abstract nature of these fundamental concepts and the difficulties involved in observing even the most basic thermal phenomena with traditional methods, an IR camera appears to be a formidable tool in addressing these challenges. However, research on the use of these devices in primary science classrooms remains limited. In particular, it would be useful to examine the ways IR cameras can support the initial stages of energy-related concept formation. Moreover, thermal cameras alone do not guarantee successful learning - further research is needed to discover how to best combine these instruments with various pedagogical approaches.

We conclude that IVB has the potential to reveal students' emotional experiences, and it offers a more diverse picture than teacher observations. Students' emotional experience does not always come across through the student's appearance. Through IVB, it is easy for students to convey their emotions to the teacher. In comparison with using paper and pen, we conclude that not all emotions can be reduced to language, and therefore also observations of facial expressions are important. If the students are familiar to the teacher, also facial expressions and gestures can be considered. Through IVB, students get the possibility to inform the teacher about their positive emotional experiences, and the teacher can better adjust the content and instructional methods to promote and maintain situational interest. However, the students also have the possibility to express their negative emotions, which can serve as a starting point to discussions. To be able to express oneself and being heard may be beneficial for the classroom interaction and students' motivation to learn science. Using smartphones instead of paper and pen is a natural means of expression for the twenty-first century adolescents.

References

Ainley, M., & Hidi, S. (2014). Interest and enjoyment. In R. Pekrun & L. Linnenbrink-Garcia (Eds.), *Handbook of emotions in education*. Routledge.

Csikszentmihalyi, M., & Hunter, J. (2003). Happiness in everyday life: The uses of experience sampling. *Journal of Happiness Studies, 4*(2), 185–199.

Dolo, G., Haglund, J., & Schönborn, K. J. (2018). Thermal cameras as a semiotic resource for inquiry in a south African township school context. *Designs for Learning, 10*(1), 123–134. https://doi.org/10.16993/dfl.96.

Finnish National Core Curriculum for Basic Education. (2014). Draft version of the new national core curriculum for basic education. *Finnish National Board of Education*. Retrieved from http://www.oph.fi/ops2016/perusteluonnokset

Haglund, J., Jeppsson, F., & Schönborn, K. J. (2016a). Taking on the heat – A narrative account of how infrared cameras invite instant inquiry. *Research in Science Education, 46*, 685–713. https://doi.org/10.1007/s11165-015-9476-8.

Haglund, J., Jeppsson, F., Hedberg, D., & Schönborn, K. J. (2015). Thermal cameras in school laboratory activities. *Physics Education, 50*(4), 424–430. https://doi.org/10.1088/0031-9120/50/4/424.

Haglund, J., Jeppsson, F., Melander, E., Pendrill, A.-M., Xie, C., & Schönborn, K. J. (2016b). Infrared cameras in science education. *Infrared Physics & Technology, 75*, 150–152. https://doi.org/10.1016/j.infrared.2015.12.009.

Hektner, J. M., Schmidt, J. A., & Csikszentmihalyi, M. (2007). *Experience sampling method: Measuring the quality of everyday life*. Sage Publishing.

Hidi, S. (2006). Interest: A unique motivational variable. *Educational Research Review, 1*, 69–82.

Hidi, S., & Renninger, A. (2006). The four-phase model of interest development. *Educational Psychologist, 41*(2), 111–127.

Juuti, K., Loukomies, A., & Lavonen, J. (2019). Interest in dialogic and non-dialogic teacher talk situations in middle school science classroom. *International Journal of Science and Mathematics Education.*https://doi.org/10.1007/s10763-019-10031-2.

Katz-Buonincontro, J., & Hektner, J. M. (2014). Using experience sampling methodology to understand how educational leadership students solve problems on the fly. *Journal of Educational Administration, 52*(3), 379–403.

Krapp, A., & Prenzel, M. (2011). Research on interest in science: Theories, methods, and findings. *International Journal of Science Education, 33*(1), 27–50.

Larsson, A., Stafstedt, M., & Schönborn, K. J. (2019). Heat angels and paper cups: Pupils' use of metaphoric relations when engaging thermal cameras to investigate heat. In E. McLoughlin, O. E. Finlayson, S. Erduran, & P. E. Childs (Eds.), *Bridging research and practice in science education* (pp. 73–89). Springer.

Litmanen, T., Lonka, K., Inkinen, M., Lipponen, L., & Hakkarainen, K. (2012). Capturing teacher students' emotional experiences in context: Does inquiry-based learning make a difference? *Instructional Science, 40*(6), 1083–1101.

Lonka, K., & Ketonen, E. (2012). How to make a lecture course an engaging learning experience? *Studies for the Learning Society, 2–3*, 1736–7107.

Loukomies, A., Juuti, K., & Lavonen, J. (2015). Investigating situational interest in primary science lessons. *International Journal of Science Education, 37*(18), 3015–3037.

O'Keefe, P., Dweck, C., & Walton, G. (2018). Implicit theories of interest: Finding your passion or developing it? *Psychological Science, 29*(10), 1653–1664.

Osborne, J., Simon, S., & Collins, S. (2003). Attitude towards science: A review of the literature and its implications. *International Journal of Science Education, 25*(9), 1049–1079.

Palmer, D. (2009). Student interest generated during an inquiry skills lesson. *Journal of Research in Science Teaching, 46*(2), 147–165. https://doi.org/10.1002/tea.20263.

Palmer, D., Dixon, J., & Archer, J. (2016). Identifying underlying causes of situational interest in a science course for preservice elementary teachers. *Science Education, 100*(6), 1039–1061.

Pekrun, R., Goetz, T., Daniels, L. M., Stupnisky, R. H., & Perry, R. P. (2010). Boredom in achievement settings: Exploring control–value antecedents and performance outcomes of a neglected emotion. *Journal of Educational Psychology, 102*(3), 531–549.

Pekrun, R., Goetz, T., Titz, W., & Perry, R. P. (2002). Academic emotions in students' self-regulated learning and achievement: A program of qualitative and quantitative research. *Educational Psychologist, 37*(2), 91–105.

Pekrun, R., Muis, K., Frenzel, A., & Goetz, T. (2018). *Emotions at school*. Routledge.

Putwain, D. W., Becker, S., Symes, W., & Pekrun, R. (2018). Reciprocal relations between students' academic enjoyment, boredom, and achievement over time. *Learning and Instruction, 54*, 73–81.

Reeve, J., & Halusic, M. (2009). How K-12 teachers can put self-determination theory principles into practice. *Theory and Research in Education, 7*(2), 145–154.

Rosebery, A., Ogonowski, M., DiSchino, M., & Warren, B. (2010). "The coat traps all your body heat": Heterogeneity as fundamental to learning. *Journal of the Learning Sciences, 19*(3), 322–357.

Schiefele, U. (1991). Interest, learning and motivation. *Educational Psychologist, 26*(3–4), 299–323.

Schönborn, K. J., Haglund, J., & Xie, C. (2014). Pupils' early explorations of thermoimaging to interpret heat and temperature. *Journal of Baltic Science Education, 13*(1), 118–132.

Shirazi, S. (2017). Student experience of school science. *International Journal of Science Education, 37*(14), 1891–1912. https://doi.org/10.1080/09500693.2017.1356943.

Short, D. B. (2012). Thermal imaging in the science classroom. *School Science Review, 94*(346), 75–78.

Silvia, P. J. (2008). Interest: The curious emotion. *Current Directions in Psychological Science, 17*, 57–60.

Vollmer, M., & Möllmann, K.-P. (2010). *Infrared thermal imaging: Fundamentals, research and applications*. Wiley-VCH.

Xie, C. (2011). Visualizing chemistry with infrared imaging. *Journal of Chemical Education, 88*(7), 881–885. https://doi.org/10.1021/ed1009656.

Xie, C., & Hazzard, E. (2011). Infrared imaging for inquiry-based learning. *The Physics Teacher, 49*, 368–372. https://doi.org/10.1119/1.3628268.

Chapter 7
A Language Model Based Analysis of Pupils' Practical Work with IR Cameras

Niclas Åhman and Fredrik Jeppsson

7.1 Introduction

Within science education research, there has historically been an extensive interest in how students and teachers comprehend different physical concepts related to the field of thermodynamics (cf. Duit, 2010). There are several reasons for having thermodynamics as a special focus of interest. For instance, heat is often described as an abstract invisible and notoriously hard concept for many students and pupils to learn. Hence, these challenges open up many exciting research interventions with the purpose of supporting pupils' and students' learning of heat; and many of these interventions often make use of visualizations and representations of different kinds (e.g., Haglund et al., 2016; Wiser & Amin, 2001).

It is well known that scientific content is mediated through a variety of resources for meaning making, such as diagrams, symbols, graphs, 3D-models, gestures, actions, and spoken or written words. Due to the ample use of models and different representations, learning science has been described as "learning to think with representation" (Klein & Kirkpatrick, 2010). For example, in relation to chemical bonds, these bonds can be described through visual diagrams highlighting the electronic structure, formulas using chemical symbols, balls and sticks representing elements and bonds, or words explaining processes and describing electronic structures. Each of these representations has its specific potential, or affordances (Gibson, 1977; Kress, 2010) for meaning making. While some aspects of the phenomenon

N. Åhman (✉)
Department of Physics and Electrical Engineering, Linnaeus University, Växjö, Sweden
e-mail: niclas.ahman@lnu.se

F. Jeppsson
Department of Behavioural Sciences and Learning (IBL), Linköping University, Linköping, Sweden

J. Haglund et al. (eds.), *Thermal Cameras in Science Education*, Innovations in Science Education and Technology 26, https://doi.org/10.1007/978-3-030-85288-7_7

will be emphasized through the choice of a specific representation, others will be de-emphasized, which is a challenge for science teaching and learning. One investigation demonstrated that different aspects of the pupils' understanding of heat conduction were highlighted depending on whether they used their drawings or their verbal explanations in the transmediating process. For example, pupils' drawings often proved to be difficult to interpret and seemingly represented a static state that is not characterized by a process. However, by letting the pupils orally elaborate on their drawings, there were several cases where the pupils managed to account for heat conduction in an adequate way (Haglund et al., 2012).

With a specific interest in supporting pupils' meaning making processes in relation to thermal phenomena, Haglund and colleagues have developed several thermal imaging (IR Camera) applications and laboratory activities across primary, secondary, higher, and informal settings (e.g. Haglund et al., 2014, 2016; Samuelsson et al., 2017). The IR camera helps to visualize physical phenomena such as friction and conduction. Visualizing heat and thermal conductivity through the use of an IR camera demystifies heat transport and has a positive impact on pupils' understanding of these concepts (Haglund et al., 2014, 2015; Xie & Hazzard, 2011). Cabello et al. (2006) have shown that the use of an IR camera facilitates pupils' understanding, and tends to make them become more motivated in the subject. Also findings by (Haglund et al., 2015) indicate that the IR camera has a positive effect by guiding the focus of pupils' attention to the intended object of study. The IR camera may also serve as a shared point of attention (Schoultz et al., 2001) in pupils' dialogues, when discussing invisible phenomena such as heat conduction and heat transfer (Jeppsson et al. 2017).

The data and analysis in this chapter will be theoretically framed within a broad perspective on social semiotics (Halliday, 1978) with a multimodal perspective of meaning making (Jewitt, 2009).

The interaction between the pupils, the teacher and the IR camera is described and analyzed in regard to Systemic Functional Linguistics (SFL), where the representations from the IR-camera are considered in terms of the three dimensions of interaction, or contexts of a situation, namely the *field* (what is going on in the situation?), *tenor* (who is taking part in the situation?), and *mode* (what role is being played by language and other ways of communication in the situation?) (Halliday, 1978; Halliday & Matthiessen, 2004). These three concepts relate specifically to the ideational (field), interpersonal (tenor), and textual (mode*) metafunction* (Halliday, 1978; Halliday & Matthiessen, 2004).[1]

[1] Grounded in Halliday's (1978) three metafunctions and with a particular focus on variables for practice discourse (Bergh Nestlog, 2012), we have published a study in the *International Journal of Science Education* (Åhman & Jeppsson, 2020), building on the same data sample – used in this chapter – and there we unpack our analyses more in detail in relation to the theory of social semiotics and speech functions in the form of practice voice, practice field and practice form.

Within social semiotics, *Tenor* – corresponding to the interpersonal metafunction – targets questions such as who is taking part in the situation, and guides our analytical focus on the relationships between the teacher and the pupils in their dialogue on thermal concepts. In this study we zoom in on, and explicitly focus on, the relationship in dialogue between the pupils and the teacher, where the IR camera serves as a shared object of attention. As a way of capturing the teacher's and the pupil's ideas of what is going on in terms of content-specific questions, we adopt the fundamental parts of Halliday's (1978) ideational metafunction. However, we do not intend to outline a complete linguistic technical analysis of linguistic items in detail, and hence we have chosen to adopt *Field* as a way to systematically analyze and describe what content is expressed in a more comprehensible and straightforward way. In this way we seek answers to the teacher's and pupils' meaning making with regard to heat conduction and insulation, in relation to the field of science teaching practice at a primary level. The third part of our analysis is guided towards how meaning (pupils' and teacher) is organised and what *Mode* of expression (gestures, verbal language, image etc.) are in play in the meaning making process.

Adopting the idea that Field, Tenor and Mode can be regarded as a kind of register used in a social context (genre) that shapes language (Thwaite et al., 2020), we aim to describe the interplay between *Field*, *Tenor* and *Mode* as a way to highlight the affordance of an IR camera within a primary science context (genre), and we focus on how an IR-image – a shared object of attention – structures the dialogue between the pupils and the teacher. Our ambition is to provide the community of science education and research area with knowledge about central aspects of meaning making in science classrooms which can be used to improve the ways in which representations are used in classroom activities. This knowledge can be further used to enhance our understanding of the fundamental processes of remaking meaning across modes (transduction) (Kress, 1977) – the procedure of translating a work into a different mode – in teaching and learning in general.

7.2 Method and Data Collection

In this study[2] we present data that involves one teacher's work with thermal phenomena in Grade 6 (12–13 year old pupils) with 45 pupils, divided into two or four groups. During teaching-led lessons the pupils were divided into two groups, and in laboratory work the pupils were divided into four groups. In the results section we have given fictive names to pupils and the teacher.

Since the teacher had no experience of working with an IR camera, we set up a workshop where the teacher conducted various experiments to explore the IR camera's functions together with one of the authors. The idea of this workshop was

[2] The data set in this study is part of a larger research project (PhD-project), Teachers' professional development in science education, led by Linnaeus University.

Table 7.1 Overview of lessons, participating pupils and method of data collection

Activity	Group	Data collection
Workshop	Researcher and teacher	Audio recording
Lesson 1	45 pupils in four groups	Video and audio recording (four groups)
Lesson 2	45 pupils in four groups	Video and audio recording (four groups)
Lesson 3	45 pupils in two groups	Video and audio recording (one group)

partly to let the teacher become familiar with the IR camera, but we also wanted the teacher to have a feeling of ownership of the experiments that the pupils eventually should work with. However, based on this workshop, the teacher chose to use the following six experiments: how ice melts while in contact with various materials such as paper and china plate; conductivity through paper and copper strips; heat radiation on black and white paper; friction of an eraser on a table and dropping a ball on a floor; how to conserve heat in a thermos; and the insulating properties of a quilted jacket.

After the workshop, the teacher planned three separate lessons for the pupils (see Table 7.1).

During the first lesson, the pupils constructed a thermos. For the thermos construction experiment, the pupils used a 0.5 litre PET bottle, and based on their ideas they could use materials such as styrofoam, textiles, tape, cotton wool and tinfoil in order to insulate the thermos. During Lesson 2, the pupils worked with the six selected experiments, using the IR camera. All the experiments involved a predict, observe and explain (POE) approach (White & Gunstone, 1992). The third lesson was a teacher-led lesson with a summary discussion and conclusions of the experiments. From Lesson 3 there was only a recording from one group, which affected which data were used from Lesson 2. Overall, in this study, we describe findings from two of the groups (25 pupils). We have collected our data by use of video and audio recordings from Lesson 2 and Lesson 3.

Between Lesson 1 and Lesson 2 the teacher handed out laboratory guides to the pupils, and the pupils wrote down their own predictions in relation to the experiments. This particular occasion was not video-recorded.

In all the involved experiments the pupil used a *FlirOne* IR camera. This particular IR camera is attached to an iPad.

The data collection was mainly done through video and audio recordings from Lesson 2 and Lesson 3. The collected data material was processed through a qualitative data analysis program of the type MAXQDA, and the analysis was performed based on the Swedish transcript by the first author. Examples provided to illustrate the results have been translated into English. Subsequently in the analysis process, key events were selected in relation to Tenor, Field and Mode. Thereafter, the first round of the analysis was discussed with other researchers with a special research interest in literacy, multimodality and learning practices. Finally, the method and our first round of analysis were presented and discussed at different seminars at our department, and at different national science education conferences.

In the second phase of the analysis, we focused on the comments we got from colleagues and we spent more time investigating how the image from the IR camera supports the pupils in their meaning making process together with the teacher.

7.3 Results

We structure our results section in terms of Tenor, Field and Modes and the excerpt represents teacher – student and teacher – teacher interaction patterns. Based on our analysis we then discuss how these registers – built up by Tenor, Field and Modes – work in conjunction with each other in order to structure a dialogue centered around the IR-camera. First, under Tenor, we show examples of typical interaction and communication patterns found in our analysis. Second, under Field, we unfold the communication in terms of reflecting on content-specific questions. Third, under Mode, we show how semiotic resources work in combination with each other naturally, as a way for the teacher and the pupils to make meaning of the experiments. The data we present here are intended to give examples of often typical pedagogical patterns that take place in many science classrooms; and we wish this contribution to be seen as a case of how a functional linguistic approach can make classroom discourse visible.

7.3.1 Tenor

Within this category we show examples of interaction patterns that were identified in our analysis. The exchange structure in the communication process between the teacher, pupils, and the IR camera was typically in the form of the teacher (Patricia) using open and closed questions together with statements as a way to guide the pupils and help them to put words on their laboratory experiences, as seen in the example in Excerpts 7.1, 7.2, and 7.3.

Excerpt 7.1: Patricia Asking Open Questions
Patricia: When you put your fingers on these two (teacher pointing to different materials), what did you experience? How did it feel?

Excerpt 7.2: Patricia Asking Closed Questions
Patricia: Was there any difference when you touched them?

In Excerpt 7.1 we see how Patricia uses open questions as a way to guide the pupil's attention towards the experiment with copper and paper strips. The rhetorical function of guiding the pupil's interest is also seen in Excerpt 7.2, albeit with the use of a closed question. These two questions explicitly refer to the pupils' embodied physical experiences of something 'being cold'. In addition to the use of open and closed questions offered by the teacher, one could also see this as an example of an

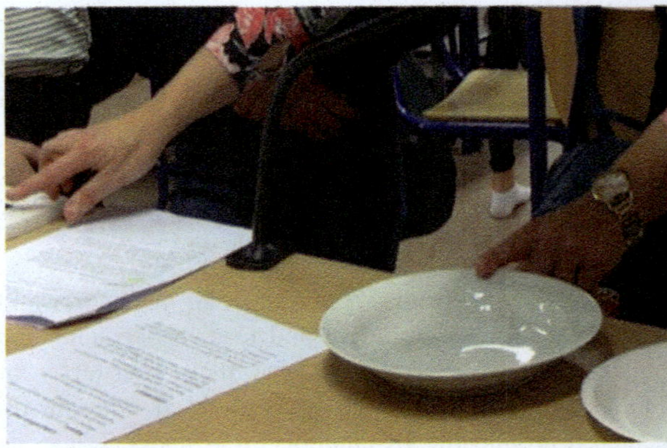

Fig. 7.1 Patricia links the two experiments regarding heat conduction

instance where the teacher is interacting with a scientific object (copper and paper strips as laboratory equipment) and how this interaction rhetorically frames the situation in a way where the pupils are supposed to connect the scientific experiment with their embodied experiences, and the pupils are left with their physical senses and their verbal language as the 'only' semiotic resources to make meaning of the situation.

Next, we will show an example where Patricia takes on the role of the influential teacher probing the pupils (see Fig. 7.1) with statements together with an open question as a way of structuring the dialogue, and as a way for the pupils to construct meaning of science as a collaborative process in an experiment on heat conduction (how ice melts while in contact with various materials such as paper and a china plate).

Excerpt 7.3: Patricia Using a Statement Together with an Open Question

Patricia: This [a china plate]. If you hold it, this feels cooler, but I promise that they are the same temperature [a china plate and a paper plate].

Asta: Mmm...

Patricia: What do you think you have done with your body heat? What do you think it [the body] has done when you have held this [the china plate]?

Astrid: It emits.

Patricia: It emits. When I send my heat to it [the china plate], my finger will feel a little colder because I have given away some heat. And this is the reason why this [the china plate] feels colder. Because it conducts heat better. And the metal conducts heat better.

First, during this experiment Patricia is interacting with the laboratory equipment and she takes on the role of the authoritative teacher, where she explicitly builds up a relationship with the pupils in terms of trust, as seen in: "If you hold it, this feels cooler, but *I promise* that they are the same temperature". In this way Patricia structures the dialogue as a close interrelation between action (the pupils' experiences) and rhetorically as a way to enable – in this case – Asta to make meaning from the experiment.

Next, Patricia supports the pupils in terms of asking more or less leading open questions, as seen in: "What do you think you have done with your body heat? What do you think it [the body] has done when you have held this [the china plate]? Here Astrid experiences a situation where she feels that the china plate takes heat from her finger and sends it out to the plate. This is confirmed by Patricia in her confirmation of Astrid's answer, "it emits" in terms of putting words on Astrid's experience as seen in: "I have *given away* some heat".

7.3.2 *Field*

Now we turn our focus to what is going on in the situation in relation to subject-specific content. As seen in other studies (e.g. Haglund et al., 2016) the pupils almost immediately started to explore things around them with the IR camera. A general impression is that the pupils, seemingly, simply related different colours to the temperature scale on the IR-camera image, and this knowledge about the colours of the image is sufficient for the pupils to make meaning and interpret the IR-image in relation to the experiments.

We will now return to the experiment on heat conduction in paper and copper strips (Excerpt 7.1). However, now Patricia is talking to two pupils when they are trying to figure out and make meaning from the experiment. In the dialogues with the pupils, Patricia is now guiding the pupils' focus towards a shared object of attention in terms of the image of the IR camera (see Fig. 7.2).

Excerpt 7.4

Patricia: It is like this. Which [paper or copper strip] is conducting heat best? Can you see [guiding Asta's attention to the IR-camera image (as seen in the photo to the right)] this [copper strip] becoming warmer? She [a pupil] gives away heat from her body to this [copper strip].

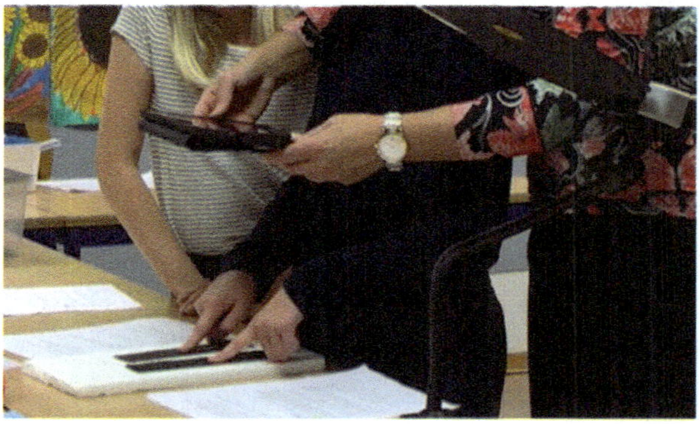

Fig. 7.2 Patricia guiding pupils' attention

Patricia rhetorically frames this situation (Excerpt 7.4) in a figurative way, where heat is being construed as some kind of substance-like entity[3] that can move from a place with high temperature to a place with lower temperature. Here, Patricia makes sure that the pupils can see this otherwise invisible process of heat conduction in the image of the IR camera. The IR camera image offered the pupils instant feedback and a visual explanation of what was happening in terms of heat conduction in the experiment, and in Fig. 7.4 we show examples of IR-camera images that were taken during the experiment with paper and copper strips. In addition, to confirm subject-specific ideas for pupils in this science class, Patricia also took the opportunity to explain heat conduction to a curious colleague who was visiting the class.

In Excerpt 7.5 Patricia is demonstrating how the IR camera works for her colleague. The colleague is looking at the paper and copper strip and can see how the IR camera image is colored yellow where the fingers are placed. Patricia then explains the subject content of the experiment.

Excerpt 7.5

Patricia: The heat from my body of 37 degrees goes more easily into the metal [strip]. Going away. So, I give away my heat to the metal. Do you understand? This is shown quite clearly.

Once again, we see an example of a situation where Patricia is modelling her verbal language – by the use of the figurative phrases "goes more easily", "going away" and "give away" – as a way to explain what is depicted in the IR camera image. As seen in Fig. 7.4, our interpretation is that the heat flow model, where heat is visualized as a yellow and red flow that actually moves in the image, frames the situation. Furthermore, the affordance of the IR camera image allows Patricia to combine her verbal language in a figurative way together with the IR camera image in a pedagogical and coherent stretch of reasoning, in a way that enables the pupils and her colleague to make meaning on heat conduction (Fig. 7.3).

The subject content in all six experiments involved heat transfer in terms of radiation and conduction. In the experiment "How to conserve heat in a thermos", the aim was also for the pupils to experience insulation and heat reflection. Some of these thermoses which the pupils made were covered by tinfoil, which for the pupils created an unexpected IR camera image. Due to the low emissivity of tinfoil, it looked like the thermos had a much lower temperature compared to areas that were not covered in tinfoil (see the blue areas on the left image in Fig. 7.4). In the discussion with the pupils during the experiment, the teachers told them that shiny metal, such as our example of tinfoil, reflects heat and makes the IR camera image incorrect.

The teacher-led third lesson had a clear content focus, and the teacher encouraged the pupils to use scientific language in their explanations of the experiments.

[3] We make this claim on the basis where Patricia construes heat (an abstract entity) in terms of a concrete object that can be given away. For an in-depth reading on the use of implicit metaphors within a science context please see Jeppsson et al. (2013).

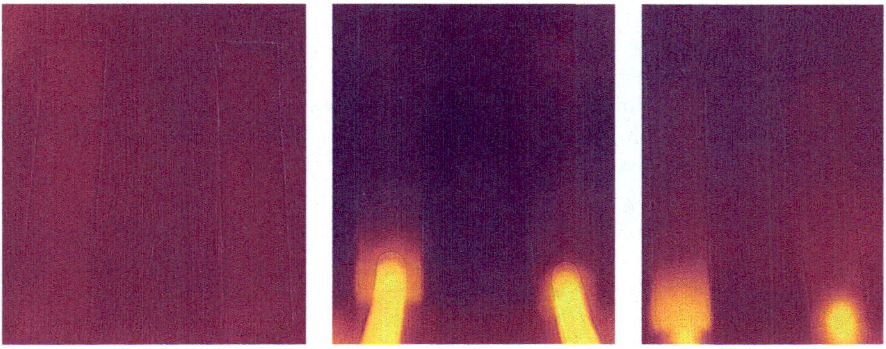

Fig. 7.3 Left image: copper and paper strips before placing fingers on them. Middle image: copper and paper strips when the pupil has placed their fingers on them. Right image: copper and paper strips when the pupil has taken away their fingers

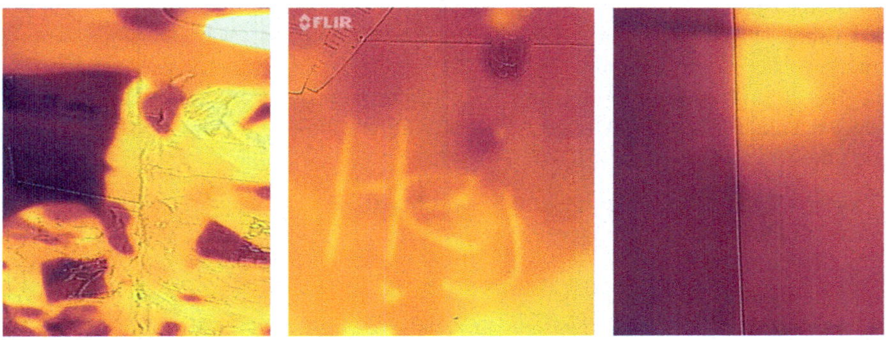

Fig. 7.4 Left image: a thermos made by pupils. The yellow stripes are tape and the blue/purple parts are covered with tinfoil. Middle image: a pupil writes the letter H on a desk with an eraser. Right image: heat radiation on white and black paper. White paper is to the left in the image

In Excerpt 7.6 we will show an example where Patricia (the teacher) – by the use of the instant visual feedback from the IR camera image – encourages Ethan and Miriam to express words about how they are working with reflection and absorption (right image in Fig. 7.4).

Excerpt 7.6: Dialogue with Content Focus

Ethan: The black paper. It attracts the light [visualised by becoming more yellow than the white paper in the IR camera image], while the white paper puts it away.

Patricia: Very good, Ethan. If we say it once more and then use some scientific words. What is it called when black paper attracts the sunlight? What is it called, Miriam?

Miriam: Absorb.

Patricia: It absorbs. And what is it called when the white ... When it bounces out again? Ian [the teacher wants Ian's attention].

Ian: Reflects.

Due to the IR camera's potential for visualizing invisible phenomena, such as heat reflection of shiny metal (Fig. 7.4, left image), it is possible for the teacher to explicitly play on the affordance of the image as a resource for pupils' meaning making process of thermal physical concepts. In Lesson 3, the teacher used pupils' experience of reflection of shiny metal to explain and sketch the construction of a thermos on the whiteboard.

7.3.3 Mode

Turning our focus towards the third lens in our analytical instrument we will give examples of the meaning making role and function that different semiotic resources play in science learning on thermal phenomena. Overall, the typical pattern we found in our analysis was that deictic gestures – a gesture that serves as a "pointer" to direct a person's attention towards a specific referent – was used most often and coordinated with more or less figurative verbal language as a way for pupils to make meaning of the experiments.

Our first illustration is an example of instant inquiry (Haglund et al., 2016) and instant visual feedback, where Elvis and Eskil are playing around with an ice cube and the IR camera (Fig. 7.5).

Excerpt 7.7: Instant Inquiry and Visual Feedback

Elvis: Oh, it is minus degrees (Elvis is really enthusiastic and is drawing a picture with the ice cube on the plate and discovers that this is visible in the IR camera image). Eskil come! Eskil!

Elvis: It's really cool. There are purple stripes [on the china plate]. Eskil, look! Look at the camera [two blue arrows have been inserted in the picture to show how Elvis and Eskil looked at the screen].

Eskil: You can write again. May I write?

Elvis: Wait, I'll write something. I want to write something . . . I'll write . . .

The affordance of the IR camera image seems to frame the experiment in a way that leads Elvis and Eskil into a feeling of ownership in their meaning making process, and in this situation the representation (the IR camera image) provides access to a disciplinary content (cf. Fredlund et al.'s (2012) discussion on disciplinary affordance). The dynamic process of meaning making that occurs in this situation calls for an analysis by a researcher or an evaluation by a teacher that stretches beyond verbal language.

Another example of coordination between the verbal linguistic mode and the visual graphical mode as an expression of instant inquiry is seen in the example where Anna, Moa and Rosa are investigating the look of themselves through the IR-camera as seen in Fig. 7.6.

Fig. 7.5 Instant visual feedback helps Eskil and Elvis to focus on temperature change

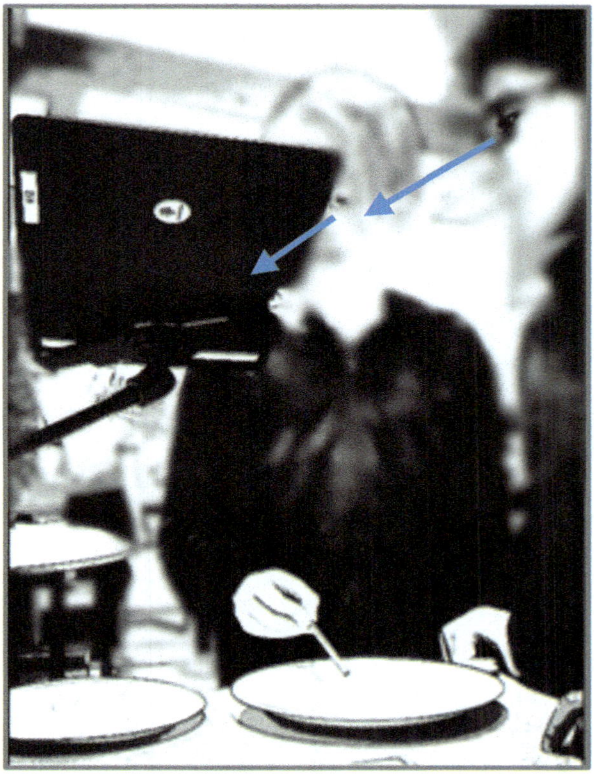

[Anna and Moa stare on Rosa through the IR-Camera and giggle]

Anna: What fun! [Points on the IR-Camera image (as seen in upper image of Fig. 7.7)]
Moa: Your glasses!
[Rosa takes the camera and studies Moa]
Moa: Down here is the camera. Down here is the camera.
Anna: The camera is down here [Pointing at the IR camera on the Ipad].
Rosa: You're quite warm here [Rosa pats Moa's cheeks]. Let's go!
Moa: Just wait. We'll see if I have a fever. Check now if I have a fever.
[turns the camera towards the forehead according to the bottom image in Fig. 7.7]. Do I have a fever?
Anna: Yes, yes, you really really do!

Here we can see how Anna and Moa play around with the IR camera, and make meaning of how the IR camera works. By looking at the images of the IR camera, they also have discussions about the meaning of the different colors. The pupils also use hand movements in their discussions, as pointing at the image of the IR camera or, based on the image on the IR camera, showing Moa where her face is warmest.

Fig. 7.6 Anna, Moa and Rosa looking at each other through the IR camera

Fig. 7.7 Patricia uses deictic gestures coordinated with verbal language and the visual graphical mode of the IR-camera image as a way to guide two pupils and the curious colleague [the man on Patricia's right hand side]

Our last example shows how Patricia uses a deictic gesture coordinated with verbal language and the visual graphical mode of the IR-camera image as a way to guide her colleague, Laura, Lucy and Mary, in a way that 'forces' them to focus on the IR camera image as seen in Fig. 7.7. In this way Patricia manages to frame the discussion in relation to both reflection and heat conduction in a coherent way that supports Laura, Lucy and Mary (and perhaps also her colleague) in their meaning making process.

Patricia: Then we made them black, because metal reflects otherwise. Now that she's looking up here [pointing at the IR-camera image according to the left picture in Fig. 7.7]. Then she (referring to Lucy that is holding the IR camera) will see that heat disappears … it will go easier in one material.

Colleague: Okay

Patricia: I do not know if you see it girls? … Do you see that it goes smoother in a material?

Laura: yes

Patricia: That the yellow disappears further away? [Point to the right picture in Fig. 7.7] What material is it?

Laura: That's it, I think [Laura points to the screen]

Patricia: What material is it, Laura?

Laura: Metal

Patricia: Metal. Metal conducts heat better. Have you done that with the ice cube too?

7.4 Discussion

In the results section we have provided examples of a possible way of using a language-based model in relation to Field, Tenor and Mode to analyze science teaching and learning in a classroom context with primary pupils. Our ambition was to examine how teachers and pupils coordinate a range of semiotic resources in their moment-to-moment interaction when they interact with an IR-camera. Just as in any other theoretical frameworks, the analytical model we have used here also has its limitations. However, there are benefits too. For instance, by having an explicit focus on Patricia's actions in terms of her social role we were able to identify language patterns (statements, open- and closed questions) that required the pupils to have a sustained focus on the experiment. Allowing oneself to focus on well-defined elements (e.g. Field, Tenor and Mode) gives one an opportunity to see patterns that may not always be visible in a wider context. In our case, we chose to isolate events in relation to a content in focus of what was going on, the structure of the interaction between the teacher and the pupils and the role played by language and other semiotic resources. Next in the discussion we will focus on the educational implications, based on how the IR image – a shared object of attention – structures the dialogue between the pupils and the teacher.

7.4.1 How Can an IR-Image – A Shared Object of Attention – Structure the Dialogue Between the Pupils and the Teacher?

Based on the examples we have highlighted in this study, we would argue that the visual and graphical modes of an IR camera image function as a shared object of attention in terms of structuring the dialogue with a sustained focus on the thermal phenomena in focus in the following way. First, the affordance of the IR camera in terms of visualizing invisible phenomena opens up a feeling of ownership for the pupils. Whether this is a feeling of actual ownership of their meaning making process or if this is another example of pupils showing curiosity in a technical gadget, the IR camera seems to have a positive effect on capturing the pupils' focus in relation to the experiments. Other studies have shown that the IR camera demystifies concepts such as heat transport (e.g. Haglund et al., 2014, 2015) and our interpretation of the dialogues in this study is that the visual graphical mode of the IR camera image functions as a resource for Patricia when she uses statements or questions of different kinds (e.g. Gibson, 1977; Kress, 2010). Hence, the coordination of verbal language and gestures from Patricia, together with the IR camera image, structures the dialogue in a step-by-step supportive way for the pupils.

Second, as part of the core content in Physics in Years 4–6, pupils are supposed to develop an understanding that "energy flows between objects with different temperatures: how the flow of energy can be affected by such means as clothes, thermos and house insulation." (Skolverket, 2018, p. 179). As seen in our study and elsewhere, the IR camera has the potential of being a truly pedagogical resource for pupils' meaning making about thermal phenomena.

Furthermore, the use of laboratory work with the IR camera also gives the pupils an opportunity to develop skills in relation to using digital equipment, as has been put forward as another core content by the Swedish National Agency for Education. If we compare this with Klein and Kirkpatrick's (2010, p. 88) idea that "learning to think with representation" is a central part of learning science, the IR camera image seems to open up a dialogue that centers around a representation and at the same time gives the pupils an experince of using modern technology in school laboratory work.

Our ambition was to provide the community of science education research and the practise area with knowledge about central aspects of meaning making in science classrooms, as a way to enhance our understanding of the fundamental processes of transmediation in teaching and learning. We think that our results have shown the potential of letting pupils express their meaning of abstract phenomena through the lens of a visual graphical interface that helps them to place words on abstract science phenomena in a coherent and adequate way, in relation to core content for Physics in Years 4–6.

References

Bergh Nestlog, E. (2012). *Var är meningen? elevtexter och undervisningspraktiker* [Where is the meaning? Pupils' texts and teaching practices], Doctoral dissertation. Linnaeus University, Växjö.

Cabello, R., Navarro-Esbri, J., Llopis, R., & Torrella, E. (2006). Infrared thermography as a useful tool to improve learning in heat transfer related subjects. *International Journal of Engineering Eduction, 2*, 373–380.

Duit, R. (2010). *Bibliography STCSE, Students' and teachers' conceptions and science education*. Retrieved http://www.ipn.uni-kiel.de/aktuell/stcse.

Fredlund, T., Airey, J., & Linder, C. (2012). Exploring the role of physics representations: An illustrative example from students sharing knowledge about refraction. *European Journal of Physics, 33*(3), 657–666. https://doi.org/10.1088/0143-0807/33/3/657.

Gibson, J. J. (1977). *The theory of affordances*. Hillsdale: Lawrence Erlbaum Associates.

Haglund, J., Jeppsson, F., & Andersson, J. (2012). Young children's analogical reasoning in science domains. *Science Education, 96*(4), 725–756.

Haglund, J., Jeppsson, F., Hedberg, D., Xie, C., & Schönborn, K. (2014). *Värmekameror gör det osynliga synligt* [Infrared cameras makes the invisible visible]. Venue.

Haglund, J., Jeppsson, F., & Schönborn, K. (2015). Värmekameror hjälper elever se naturfenomen [Infrared cameras help pupils to see science phenomena]. *Lmnt-Nytt, 1*, 18–21.

Haglund, J., Jeppsson, F., & Schönborn, K. J. (2016). Taking on the heat – A narrative account of how infrared cameras invite instant inquiry. *Research in Science Education, 46*(5), 685–713. https://doi.org/10.1007/s11165-015-9476-8.

Halliday, M. A. K. (1978). *Language as social semiotic. The social interpretation of language and meaning*. Edward Arnold.

Halliday, M. A. K., & Matthiessen, C. (2004). *An introduction to functional grammar* (3rd ed.). Routledge.

Jeppsson, F., Haglund, J., Amin, T. G., & Strömdahl, H. (2013). Exploring the use of conceptual metaphors in solving problems on entropy. *The Journal of the Learning Sciences, 22*(1), 70–120. https://doi.org/10.1080/10508406.2012.691926.

Jeppsson, F., Frejd, J., & Lundmark, F. (2017). "Wow, it turned out red! First, a little yellow, and then red!" 1st-graders' work with an infrared camera. *Journal of Research in Childhood Education, 31*(4), 581–596. https://doi.org/10.1080/02568543.2017.1347589.

Jewitt, C. (2009). *The Routledge handbook of multimodal analysis*. Routledge.

Klein, P. D., & Kirkpatrick, L. C. (2010). Multimodal literacies in science: Currency, coherence and focus. *Research in Science Education, 40*(1), 87–92. https://doi.org/10.1007/s11165-009-9159-4.

Kress, G. (1977). Tense as modality. *UEA Papers in Linguistics, 5*, 40–50.

Kress, G. (2010). The profound shift in digital literacies. In J. Gillen & D. Barton (Eds.), *Digital literacy: A research briefing by the technology enhanced learning phase of the teaching and learning research programme*. London Knowledge Lab.

Samuelsson, C. R., Haglund, J., & Elmgren, M. (2017). *Looking for solutions: University chemistry and physics students interacting with infrared cameras*. Paper presented at the 12th Conference of the European Science Education Research Association (ESERA), Dublin.

Schoultz, J., Säljö, R., & Wyndhamn, J. (2001). Heavenly talk: Discourse, artifacts, and children's understanding of elementary astronomy. *Human Development, 44*(2–3), 103–118.

Skolverket. (2018). *Curriculum for the compulsory school, preschool class and school-age educare 2011: Revised 2018*. Skolverket.

Thwaite, A., Jones, P., & Simpson, A. (2020). Enacting dialogic pedagogy in primary literacy classrooms: Insights from systemic functional linguistics. *The Australian Journal of Language and Literacy, 43*(1), 33–46.

White, R., & Gunstone, R. (1992). *Probing understanding*. Falmer Press.

Wiser, M., & Amin, T. (2001). "Is heat hot?" inducing conceptual change by integrating everyday and scientific perspectives on thermal phenomena. *Learning and Instruction, 11*(4), 331–355.
Xie, C., & Hazzard, E. (2011). Infrared imaging for inquiry-based learning. *Physics Teacher, 49*(6), 368–372.
Åhman, N., & Jeppsson, F. (2020). Teachers' and pupils' scientific dialogue in learning about invisible thermal phenomena. *International Journal of Science Education, 42*(18), 3116–3133. https://doi.org/10.1080/09500693.2020.1852334.

Chapter 8
Upper-Secondary Students' Use of Infrared Cameras in the Study of Animals' Temperature

Jesper Haglund, Sofia Folestam, and Anna Westlund

8.1 Introduction

Educational development and research on how infrared (IR) cameras can be used in science education has, to this point, primarily focused on teaching in physics and chemistry (Samuelsson, Chap. 4). However, due to the versatility of the technology, there are also opportunities in the teaching of other subjects, including biology. This chapter was written as part of a joint research and development project in collaboration between upper-secondary teachers at the Swedish Natural Resource Use programme and an educational researcher, where we explored the opportunities to introduce IR cameras in teaching about animals' physiology. The aim of the study was to provide insight into how students may use infrared cameras in inquiry-based investigation of animals' temperature, and how the technology can support their understanding of animals' thermal regulation.

8.1.1 Teaching and Learning of Animals' Temperature Regulation as a Complex Phenomenon

Educational research has found that many students struggle with understanding complex systems, as they tend to focus on perceptually available structures, but fail

J. Haglund (✉)
Department of Engineering and Physics, Karlstad University, Karlstad, Sweden
e-mail: jesper.haglund@kau.se

S. Folestam · A. Westlund
Lillerudsgymnasiet, Vålberg, Sweden
e-mail: sofia.folestam@lillerud.se; anna.westlund@lillerud.se

J. Haglund et al. (eds.), *Thermal Cameras in Science Education*, Innovations in Science Education and Technology 26, https://doi.org/10.1007/978-3-030-85288-7_8

to recognise invisible, dynamic phenomena (Hmelo-Silver & Pfeffer, 2004). Complex systems are characterised by multiple levels of organisation and interrelations with other systems, and dynamic relationships between these levels and systems that are often not obvious. In their study of experts' and novices understanding of aquaria, Hmelo-Silver and Pfeffer found that while experts integrated functional and behavioural elements with structural elements, such as commenting that rocks in the aquarium have the function to facilitate breeding of fish, novices were confined to identifying structure, e.g. merely noticing that there are rocks in the aquarium.

Humans' and animals' temperature regulation is a good example of a complex phenomenon, as it involves dynamic interaction of different mechanisms and adaptations. Animals' temperature regulation involves the maintenance of their body temperature within narrow limits, also during changing external conditions. This is done by a range of different adaptations and mechanisms, including thermal insulation, metabolic heat production, circulatory adaptations, such as controlling the amount of blood that flows to different body parts and close to the skin at different external temperatures, and evaporative cooling, such as sweating or panting (Campbell et al., 2008).

Not surprisingly, educational research has identified that many students have challenges in grasping temperature regulation and related topics. Simpson and Marek (1988) provided upper-secondary students with a scenario where a person undergoes a physical fitness examination in order to assess their understanding of homeostasis, the dynamic steady state of a well-functioning body. However, none of the students was able to express a sound understanding of the mechanisms for how the person could have an unchanged body temperature after running 500 yards, in which the release of energy due to cellular respiration could be balanced by evaporative cooling as the person started to sweat. Barak et al. (1999) studied grade 10 students' understanding of animals' temperature regulation, from written responses to the following question: "Why is the temperature of a polar bear maintained constant, as opposed to that of a metal cube, when left in the snow"? The students' explanations were categorised in terms of a focus on matter or processes (Chi et al., 1994). Students with a matter focus attended to concrete features, such as the fat and fur of the polar bear that provides insulation. Only a minority of students gave process-oriented answers, where they accounted for the role of metabolic processes in balancing heat losses in living organisms' temperature regulation, interpreted as a more meaningful level of understanding.

8.1.2 Applications of Infrared Cameras in Biology and Biology Education

There are many different biology-related applications of IR cameras, including in the study of plant physiology (Jones, 2004), human medicine (Lahiri et al., 2012), and veterinary medicine (Soroko & Howell, 2018). In particular, with decreasing prices, simple handheld IR cameras can be useful for farmers and animal caretakers in different ways (Gilbertsson et al., 2019). For example, they can be used to identify animals' increased temperatures due to injuries or infections, and temperature increase in silage, which indicates unwanted biochemical reactions that reduce the quality of the fodder.

Haglund and Schönborn (2019) and Wong and Subramaniam (2020) propose a number of ways in which IR cameras can be used in biology education, relating to topics such as human, animal and plant physiology. In their analysis of how cooling and heating of bodies depends on the surface-to-volume ratio, Planinšič and Vollmer (2008) use IR images of cheese cubes in a microwave oven to model the heating process, and draw parallels to thermal regulation of land-based mammals of different sizes. In practice, Leguillon (2019) shows how many different phenomena relating to the surface temperature of animals and plants, including exposure to sun light and cooling by evaporation, can be studied with an IR camera at a pet zoo and its surrounding woods in the summer.

Unger et al. (2020) developed a laboratory activity where college biology students posed hypotheses about and investigated the temperature of endothermic (e.g. mammals and birds) and ectothermic (e.g. reptiles) animals with a thermal camera attached to a smart phone. Students found that an anole lizard placed under a heat lamp had a higher temperature than the surrounding air, and concluded that calling such ectothermic animals "cold-blooded" is misleading. Eighty-six percent of the students agreed or strongly agreed with the statement "using the thermal imaging camera helped me further understand the difference between endothermic and ectothermic animals" after the exercise.

8.2 Method

8.2.1 Context of the Study

The study was conducted in the context of an upper-secondary school that gives the vocational Natural Resource Use programme, preparing for vocations in farming and animal care. The study was part of a research and development collaboration between eight teachers at the school and Jesper Haglund (JH), a researcher that studies the use of IR cameras in science education. The teachers designed lectures in which the students were given the opportunity to use IR cameras in relation to the curriculum, with the overarching objective of evaluating the IR camera as a pedagogical tool in their different subjects. In one of the other investigated courses, Haglund and Henriksson (2021) found IR cameras to be useful tools in students' investigation of temperature increase due to injuries in pigs, and, more surprisingly, temperature decrease in capsulated infections in a sow's udder.

Sofia Folestam (SF) teaches the course Horses 1 in grades 11–12, where students' laboratory and investigative work focuses on horses' exterior, motion, physiology, genetics, and breeding, and the central content includes horses' body temperature, breathing, and pulse. The students in focus in the present study also follow an extended study programme in natural sciences, to prepare for university studies in for example veterinary medicine.

Anna Westlund (AW) teaches the course Animal biology in grade 11, where one of the objectives is that students should be given the opportunity to develop their scientific thinking and a biological understanding of animals, through laboratory and investigative work. The central content includes animals' energy conversion, blood circulation, and temperature control.

8.2.2 *Design of Activities*

Xie and Hazzard (2011) point out the possibility to use IR cameras for inquiry-based teaching approaches, where students are allowed to play a more active role in deciding what to study and how to study it than in more traditional, cookbook-style labs. Pedaste et al. (2015) have conducted a systematic literature overview of studies of inquiry-based learning, and provide a synthesis of the different steps that students typically go through in an inquiry cycle. The steps include orientation about a topic, formulation of questions and hypotheses in relation to a phenomenon, investigation of the phenomenon, and drawing conclusions from the investigation.

In our project, we adopted an inquiry-based approach where students were divided into small groups and given the task to identify something they would like to investigate with the help of an IR camera. In some of the classes, the students were quite free to come up with relevant topics, and formulate questions and hypotheses, whereas students in other classes were provided a few topics to start from. At the end of the lessons, after the students had conducted their investigations in small groups, the involved teachers and students and the researcher gathered, and together looked at and interpreted IR-camera images from the groups projected onto a screen.

The students in Horses 1 were divided into three groups of 3–4 students. During a first lesson, they were introduced to and were acquainted with the IR cameras and given the opportunity to look at horses and other animals with the cameras without much instruction. In a second lesson, the following week, SF prepared three alternative assignments for student groups. As one group of students had attended to temperature differences between the legs of horses during the first lesson, and had started to come up with possible hypotheses for why this happens, e.g. the colour of the fur of the legs influences the temperature, they were encouraged to pursue an investigation of these underlying factors.

As in the case of Horses 1, the students in Animal biology were presented to IR cameras during a first lesson, where they were also instructed to prepare an investigation in small groups guided by their own formulation of questions in relation to the studied phenomenon. During a second lesson, they conducted the investigations in and around the "Animal house", where common pets, e.g. guinea pigs and parrots, are kept, as well as some more exotic animals, such as ferrets, snakes and tortoises. Different breeds of rabbits are kept indoors, but also outdoors in cold conditions. A third lesson was dedicated to writing reports of findings and conclusions.

8.2.3 Data Collection and Analysis

Video data were recorded with handheld video cameras from selected student groups' investigations of the phenomena, and from the joint interpretations of the groups' IR images at the end of the lessons. In addition to video data, students' written reports of the investigations were collected from the class in Animal biology.

The analysis focuses on whether and how the students engage in the different steps of inquiry cycles (Pedaste et al., 2015), and their different levels of understanding of animals' temperature regulation, in terms of a focus on structural features or involved processes (Barak et al., 1999).

8.3 Results

The results of the study are provided in the form of dialogues and written reports from the investigations conducted by two groups of students, followed by analyses of the groups' ability to engage in scientific inquiry and their understanding of thermal regulation.

8.3.1 Investigation of Factors Underlying Differences in Temperature of Horses' Legs

In Horses 1, JH followed the group of students that inquired into factors behind differences in temperatures of horses' legs (see Fig. 8.1, left). In preparation, SF gave the following written instruction to the group:

> Compare dark and white legs (make sure the thickness of the fur is about the same for the horses, which it probably is if we have not clipped the legs). Compare the legs of at least four horses. All horses should be in paddocks. Measure from both sides, take pictures and notes. Discuss in the group why the measurements turned out as they did.

At the beginning of the second lesson, the students get together in class to plan their investigation. JH asks what they plan to do:

JH: What did you see the last time, regarding legs... that legs of different colour had different temperature... ?

Kate: Well, we took a regular [IR] picture of the horse... and we saw that it was a lot colder where the white stocking [leg marking] starts on

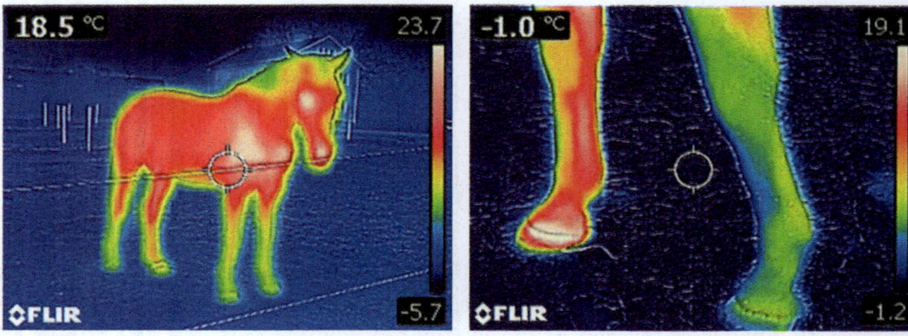

Fig. 8.1 IR image of a horse outdoors in cold conditions, showing that the legs are colder than the barrel (left); IR image of the front legs of a horse outdoors in cold conditions with a known injury in the right leg, which has a higher temperature than the left leg, and where the front parts of each leg are colder than the back parts (right)

the horse . . . and so, we measured the temperature there . . . and it was a bit different . . . and then, we wondered if it was like this for all . . . or if it was just temporarily . . . /. . ./

JH: And how could you investigate what might be the cause of this . . . did you have any ideas there . . . ?

Miriam: Well, to investigate more horses who have white legs, or white stockings . . . and check if it is like that on all of them . . . all of them are not . . . if there was an injury, or something, that made this leg colder or warmer . . . if all have it, it has something to do with the colour . . .

JH: Sofia touched upon if they differ in how thick the fur is . . . if the fur maybe is thinner where they are white . . .

Miriam: That's a thing you have to compare . . . to make sure the horses are about the same . . .

Kate: And I think . . . well, the skin colour of the horse . . . may have an influence . . . because where they are white, the skin is pink. And where they have another colour, the skin is black, most often. And that can have an influence. Possibly how long the markings are . . . we are going to check a horse that only has white fetlocks, and another one that has longer stockings that go up along the legs.

During this planning of the investigation, the students articulate the question, what they are going to investigate, and come up with a number of different factors that may influence the temperature of different parts of horses' legs, including, apart from the colour of the fur, injuries, thickness of the fur, and body part. They also use technical language that shows a knowledge of the anatomy and leg markings of horses. Next, the students compile a list of four horses they want to study, which represent a distribution of characteristics in leg colour, and draft a table to make notes of the temperatures in line with the written instruction.

Out in the paddocks, they choose to start the investigation with Charlie, Kate's horse that has a white sock on the left hind leg, while the other legs are brown, and a known injury in the front right leg that has caused lameness. Kate holds the horse, as Miriam takes still IR images on the outsides and insides of the legs in a systematic fashion at the same point on the legs, and concludes:

Miriam: I mean, you can see . . . what you can see in the IR camera, is . . . at the white [points to the left hind leg] it is green . . . and at the other side . . . [points to the right hind leg] the corresponding hoof is red, in the IR camera. So it looks as if it's colder . . .

Carolina: But it is colder on the white one?

Miriam: Yes. /. . . /

Carolina: But I still think it is a bit odd that it is colder where it is white, if the fur is thinner . . . because then it should release more heat . . . it should be the other way around.

Kate: Was the fur thinner . . . ?

Carolina: I mean, if the fur is thinner at the white . . .

Kate: Yes, that should not be right . . .

Carolina: That it should release more . . .

Kate: But I don't think it is thinner.

Carolina: No, I don't think so either . . . it is just that, if it had been so, that the fur was thinner, it would have been the other way around.

Miriam notices the pattern that the leg with white fur is colder than the leg with brown fur, and initially they conclude that a temperature difference is in line with their observations from the previous week and their expectations. However, Carolina is puzzled about the measurements, and argues that if the white fur is thinner than the brown fur, more heat should be released from the leg, and the leg would have a higher temperature in the IR camera, not lower, as is the case now. In this way, Carolina shows an understanding of the insulating function of the fur, and makes a reasonable inference from the measurements.

As they move on to the other horses, the students also start to think of how the inner anatomy of the horses' legs influences the temperature:

Kate: What I think I have seen . . . is that on the front side of the fetlocks, all horses have been a bit colder . . . compared to back at the bulbs [soft tissue just above the back of the hooves] . . . but that is because the pastern [in the back of the legs] is in view when we take the pictures [from behind] and there, the fur is much thinner . . . /. . . /

Carolina: But it may be that it is longer to the blood vessels . . . the blood vessels are at the back, and then it gets colder . . . that is why it is warmer at the back side.

After the students have taken IR images of the four horses, they go back to the classroom and go through the images, note the inside and outside temperature on the four legs of all horses in their notebooks, and search for patterns. In addition, Kate attends to Charlie's lameness:

Kate: So this one, for example . . . [shows IR image in Fig. 8.1, right] These are brown legs, both of them . . . But this horse is lame in that leg [points to the warmer right leg] and that is quite visible, right . . . because that [points to the colder left leg] gets very cold in comparison . . . so we have to ignore that . . . and, as I said, the legs were brown, so that was not very . . .

JH: But I still think it was interesting that there was such a large difference . . .

Kate: Yes, absolutely.

JH: . . . between the whole legs . . . because what I have seen before . . . if there has been a lame horse . . . only the joint has been . . .

Kate: Yes, precisely . . .

JH: . . . looked . . . has been warmer . . .

Kate: That depends on how it has been stressed . . . I mean, where the pain is . . . with this horse, it is in the hoof, under the hoof, in the sole . . . which means that as soon as he steps on something, it radiates up in the entire leg . . . [gestures motion upwards] he has been stressed a bit too hard.

JH: So, which leg was the warmest . . . ?

Kate: Right front.

JH: The injured one . . . ?

Kate: Yes, the injured one . . . /. . . / [shows another IR image of the legs] It is primarily the lower part here [points to the injured leg] that is the warmest.

JH: So, it is red there . . .

Kate: And completely white down here [points to the hoof]

JH: . . . and down towards green . . . if you compare to the other leg, that's a lot colder.

Kate: Yes, that's the normal temperature.

JH: So you could imagine that more blood is streaming to . . .

Kate: [nods] . . . for the healing process . . .

JH: . . . for the healing process . . . so that's why it's kept warm . . . but it's shut off more for the healthy one . . .

Kate: It's warmest here where it is white . . . [points to the injured hoof] and that's where he has pain.

Finally, the groups gather in the classroom to report back what they have found during their investigations. Kate reports on the findings from their group, and starts by telling that Charlie's increased temperature due to the injury of the front right leg was clearly visible, but when comparing the healthy hind legs, the white hind leg had a lower temperature than the brown hind leg. The second horse, Bonnie, in contrast, had quite even temperatures of the four legs, in spite of different colours. A general pattern, however, was that the inside of the legs generally were warmer than the outsides, apart from patches that were wet or muddy. Rosie, the third horse, stood out among the four they studied:

Carolina: And what was interesting with Rosie was that the temperature of all her legs was higher than that of Charlie's injured one.
Kate: She was very warm, in comparison. The other ones were around, like, 6–7 degrees ... she was, like, 18, 14, and 17. /.../
Lisa: Did you see how many blankets she had on?
Kate: I think it was unpadded, a woollen blanket.
Miriam: She had two blankets.
Carolina: Yes, it was double.
Kate: I don't know how much difference it does for the legs ... because if they freeze in the body, they shut off the blood supply to the legs a bit more, so then they are colder.

Kate starts summing up the group's conclusion:

Kate: So, really, the only thing we found out was that all horses are warmer on the insides of the legs than the outsides ... [laughs]
SF: Where on the inside?
Kate: On the middle of the fetlock.
SF: Not up or down ... ?
Kate: Yes, that's where we measured. It was easiest in that way. And we saw that they were warmer in the pastern than on the front side of the fetlock.
JH: What was the conclusion between white and dark parts?
Kate: Well ...
Miriam: You see a small increase on white ...
Carolina: No, the other way around! [laughs]
Miriam: No, decrease, I mean! Decrease in temperature, but it's not so marked that you can ...
Kate: We would have had to look at more horses, maybe ten, healthy ones ...
Carolina: ... in dry paddocks.
Kate: ... where the entire fetlocks are covered ... so the conditions are the same ... that was a bit hard to get now ... correct results ...

Overall, the students follow steps in an inquiry cycle (Pedaste et al., 2015) in a structured and well-thought-out way:

- The study's question is based on an initial observation of their own that intrigues them.
- The students formulate a hypothesis for why legs differ in temperature: that it has to do with the colour of the fur, but also consider other possible factors that may influence.
- They come up with a procedure of how to measure temperature, where they measure the horses at exactly the same spot, the fetlock, and carry out the observations accordingly.

- As they conduct their investigation, the students also attend to details of the observation that are unexpected, such as the white sock of Charlie being colder than the corresponding brown leg.
- The students are careful when they draw conclusions, they have identified patterns that underlie temperature differences, but have not confirmed the original hypothesis.
- They discuss the weaknesses of the observations, and how the data could be made more reliable. As often, they conclude that more investigation would be needed.

In addition, the students' engagement with the assignment provides evidence of good content knowledge in the anatomy and physiology of horses. For example, Carolina expresses an understanding of the thermal insulation of fur when she reasons about the finding that a white sock is colder than the corresponding brown leg, and they explain temperature differences between front and back parts of the legs in terms of difference in blood supply. In particular, they attend to mechanisms of regulating the blood supply to the legs for keeping the bulk of the body warm and healing the wounded leg. In these aspects, the students maintain a focus on the physiology, i.e. a process perspective on, among other things, horses' temperature regulation (Barak et al., 1999; Simpson & Marek, 1988).

8.3.2 Investigation of the Temperature of Rabbits' Ears

In Animal biology, JH followed a pair of students who chose to investigate the ear temperature of rabbits of different breeds. Before the observations, they formulated the following question: "Which breed out of Flemish Giant, Netherland Dwarf, Gotland rabbit, and Miniature Lop has the warmest ears?", and a hypothesis: "We think the Netherland Dwarf has the warmest ears, because their blood vessels lie in a more compact way and therefore excrete more heat".

The students start their investigation by going into the animal house. Susanne picks up a Netherland Dwarf rabbit [a small breed with small ears] from a small rabbit hutch indoors and holds it in front of Helena, so that she can look at it with the IR camera:

Helena: [looks at the rabbit with the IR camera] Oh, they [the ears] are completely blue! How are we going to get anything out of this . . . ? / . . . / They are completely blue! [laughs]
JH: Is that the ears . . . that are completely blue?
Helena: Yes. / . . . / I thought they would be a bit . . .
Susanne: I thought they would be warm.
Helena: Yes.
Susanne: [touches the rabbit's ears] They are quite cold.
JH: What can be the cause of this . . . that they are blue . . . that they are cold . . . ?
Susanne: That they are thin . . .

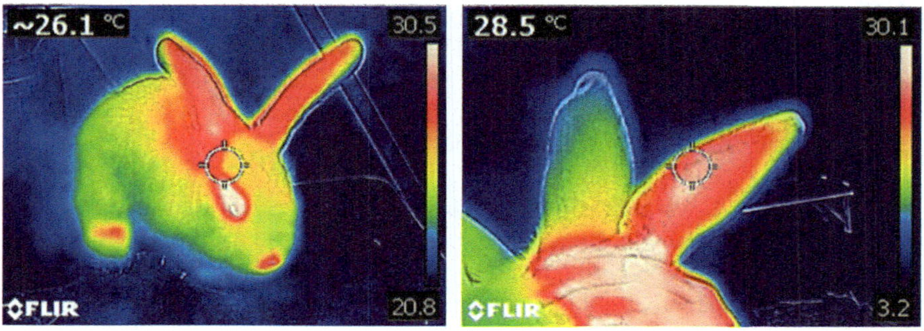

Fig. 8.2 IR image of a Flemish Giant with warm ears (left); IR image of a Gotland rabbit kept outdoors, whose right ear is warmer than the left ear (right)

Helena: Well, they are quite small. We thought of comparing with Flemish Giant Rabbit [the world's largest domestic rabbit breed] that has considerably larger ears . . . and see if they have warm ears . . . but . . .

JH: Is that your investigation . . . you have . . . ?

Helena: Yes. I mean, the breeds . . . have considerably larger . . .

Susanne: But we can look in gen . . . we don't have to do the ears . . . we can maybe look in general, on the body . . . then maybe we can check the outdoor rabbits, so we can compare three . . . first size [between the indoors rabbits] and then also the outdoor rabbits. / . . . /

Helena: The nose is quite warm, too . . .

Susanne: How about looking at the nose, then . . . ? Compare the nose outdoors and indoors . . . ?

Helena expresses surprise when she sees that the ears of the Netherland Dwarf are blue and cold, going against their prediction that the ears would be warm. They are confused about how to handle the situation, and Susanne is prepared to abandon their focus on ear temperatures and instead look at the rabbits' body temperatures in general.

They go over to a section with Flemish Giants, where Helena looks at one of the rabbits with the IR camera [see Fig. 8.2, left], and says: "They have considerably warmer ears!" Susanne picks up one of the Flemish Giants and holds it in front of Helena, who takes IR images:

JH: So, how does it differ from the small ones?

Helena: Well, their [Netherland Dwarfs] ears weren't red at all . . . anywhere . . . apart from down here where they join the head . . . but these are much redder. / . . . /

They move to another section, and Susanne picks up a Miniature Lop. Helena looks at it with the IR camera:

Helena: These [the ears] were also redder.
Susanne: So we can maybe go with the ears, then . . . as we had thought from the beginning . . . ?
JH: What did you think of the ears beforehand . . . ?
Susanne: We thought that the smallest ones had . . .
Helena: . . . warmest . . .
Susanne: . . . warmest ears . . . because the blood vessels are tighter . . .
Helena: . . . quite compact . . . our hypothesis was not correct . . .
JH: It was precisely the other way around . . .
Helena: Yes, precisely . . .
JH: And, I mean, that's always very exciting . . .
Susanne: [laughs] Yes.
JH: . . . when it doesn't turn out as one had expected.

After having looked at rabbits of three breeds, they conclude that their hypothesis of smaller rabbits having warmer ears was wrong. JH expresses enthusiasm toward this result, but the students do not share such enthusiasm. They are still puzzled about how to handle the unexpected result.

They walk out to the unheated house with outdoor Gotland rabbits. Susanne picks up a rabbit and Helena looks at it with the IR camera. She notices that the ears have different temperatures:

Helena: This [touches the right ear that looks red, see Fig. 8.2, right] has been lying against his body now . . .
Susanne: Mm . . .
Helena: So it's a matter of if that has some effect . . .
JH: So the two ears look very different.
Helena: Yes, they do. That's a bit funny, that they do . . . Why is that . . . ?
Susanne: I guess it is as you say . . . / . . . / [they touch and compare the ears]
JH: Did you feel a difference . . . ?
Susanne: Yes.

When Helena attends to the temperature difference between the two ears of the rabbit, she provides a reasonable potential explanation, that one of the ears has been heated by the rest of the body. In this way, she goes beyond structural features of the animal and focuses on the process of heating.

Next, Helena notices that the temperature is higher on the belly where the fur is thinner:

Helena: I suppose, when they are out, they are at rest a lot to keep the body temperature . . . but the question is where it is kept . . . does it remain in the body . . . then it's maybe not as warm around . . . compared to the ones that are indoors, where it is always warm . . .
Susanne: Yes, precisely.
Helena: I think a bit like that . . . since these are out . . . they maybe have thicker fur, as well, because they are out . . . then maybe the heat does not push through . . . but on the ears, it's quite thin . . .

Susanne: I don't think the ears have the same need to be warm as the rest of the body . . . I mean, in general . . .
Helena: Their ears get cold when it's very cold . . . as it's a part that sits outside . . .

In reaction to seeing the differences in temperature on different parts of the outdoor rabbits, Helena brings up aspects of their temperature regulation, such as reasoning about differences between the animals' body and skin temperatures and how thick fur helps them keep the temperature. Similarly, Susanne points to the priority of keeping the main body warm at the expense of the ears.

When they have finished their investigation, the students head back to the animal house and start drafting their report on a laptop:

Helena: Are we going to keep our question?
Susanne: I don't know . . . there was a difference between the ears . . . between all of them . . .
Helena: Mm . . . But we should maybe change . . . leave out . . . that we only compare . . . all breeds . . . that we don't write, like, only Netherland Dwarf and Flemish Giant . . . now, we have compared these breeds, and their ears . . . and also their noses, if that is worth including . . . / . . . /
JH: But, as I said . . . I thought that it was interesting that it did not turn out exactly as you had predicted . . . your hypothesis . . . then it's interesting to try to find another explanation . . . than what you had thought . . .
Helena: So of course we have to keep the hypothesis [they turn to writing the report] / . . . /
Susanne: But when we talked about animal medical care before . . . the smaller you are, the higher temperature you have . . .
Helena: Hm, that is a bit strange . . . that we saw such big differences on the ears . . . I guess it's because they are so small and don't gather as much [heat?] / . . . /

In their final report, Helena and Susanne present the results from their investigation:

> We have established that the rabbit breed with the warmest ears was the Miniature Lop (35,9 °C) / . . . / After scrutinising the images, we think the result turned out in this way because the ears of the Miniature Lop lie down, and therefore they cannot cool them down in the same way as rabbits with ears that stand up can. Also Gotland rabbits have ears that can lie down, but the ones we examined were outdoor rabbits, versus the other breeds that were indoor rabbits, and therefore their ears were colder than those of the other breeds.

They conclude that their hypothesis was wrong and provide explanations for this that incorporate aspects of the process of temperature regulation:

> It turned out that the larger breeds were warmer and not the smaller ones. That's because the larger breeds generate more heat and also consume more energy than the smaller breeds. This makes their ears warmer. / . . . / Rabbits also use their ears to cool down the rest of the

> body. By folding them out, the blood in the ears is cooled down. The cooled-down blood is then sent into the body, which cools off the body, as well. Therefore, to keep the heat, they fold down the ears towards the body. Our investigation can also have been influenced by the rabbit's current shape. If the rabbit has eaten well, more energy is generated which also represents more heat.

This pair of students are not as structured in the approach to conducting the investigation, compared to the group who studied horses' legs. The investigation is guided by a clearly stated question and a hypothesis, but when they make observations that contradict their hypothesis, they get second thoughts and start to question their approach. They take IR images of the rabbits' ears and noses, but not in a very systematic fashion. Even when JH encourages them to come up with explanations for the unexpected result, they are puzzled about how to handle the situation. However, in the final text they handed in, reasonable explanations are provided, where they have come further than during the observations, presumably after having consulted external literature or from discussions in class.

As for their understanding of animals' temperature regulation, the students' initial question and early observations indicate an interpretation of temperature as an intrinsic characteristic of an individual animal, a matter perspective on concrete features (Barak et al., 1999). Susanne backs up this view by referring to previous teaching, where they have learned that smaller animals tend to have higher temperatures than larger animals, relating to how cooling depends on the surface-to-volume ratio of physical objects as well as living organisms (Planinšič & Vollmer, 2008). Therefore, they are surprised when the first animal they observe is a small rabbit with cold ears. Further into the investigation, however, they start taking into account a process perspective on the rabbits' temperature regulation (Barak et al., 1999). Examples include when they discuss how the ears of a Gotland rabbit kept outdoors can have different temperatures and think that its heat is concentrated to the bulk of the body whereas the ears are not as vital. In the written report, they express an understanding of the ears' active role in temperature regulation, and reason about what the rabbits may have been doing before the measurement in providing explanations for their temperature, rather than merely focusing on their sizes.

8.4 Discussion

We would like to end the chapter by returning to the aim of the study, to provide insight into how students may use infrared cameras in inquiry-based investigation of animals' temperature, and how the technology can support their understanding of animals' thermal regulation.

Even though the student groups differed in ability in conducting inquiry-based tasks, the study shows that the technology can be a valuable tool in the teaching of animal anatomy and physiology in natural settings. The students in Horses 1 were more accustomed to following the steps in the inquiry cycle (Pedaste et al.,

2015). For example, they took notes in a structured way and were excited to try to develop new explanations when their predictions did not hold. In addition to a focus on understanding animals' temperature regulation, the students in Horses 1 also used the IR camera as a diagnostic device in relation to known and suspected injuries, which is an important medical application of the technology (Soroko & Howell, 2018). This initiative among the students to follow up what they wonder about is similar to what Haglund and Henriksson (2021) found in another class at the school where students noticed a temperature decrease in capsulated udder infections. This indicates engagement and a commitment to finding answers to their genuine questions, an important aspect of inquiry-based teaching. In contrast, the students in Animal biology struggled with following the steps of the inquiry cycle, and considered to abandon the question they formulated before the investigation when they could not confirm their predictions. However, in their report they were able to provide reasonable explanations for why their initial predictions were not correct, which we take as evidence of insight into the process of inquiry.

The students attended to many details in their investigations, such as how the horses' legs tended to be warmer on the front sides than the backsides and that an individual rabbit's ears could have different temperatures. In line with previous research on upper-secondary students' use of IR cameras in laboratory exercises (Haglund et al., 2015), they retained a focus on relevant, thermal aspects of the studied phenomena, which reinforces the engagement with the activities. From a science content perspective, as opposed to previous studies (Barak et al., 1999; Simpson & Marek, 1988), the students in the present study adopted a process perspective on temperature regulation, either, as in the case of the students in Horses 1, throughout the activity, or, represented by the pair of students in Animal biology, gradually moving from a matter to a process focus when conducting and reporting on the activity. In this way, the study provides an example of how the holistic, dynamic images of an IR camera can help encourage a process perspective on complex science phenomena such as animals' thermal regulation.

References

Barak, J., Sheva, B., Gorodetsky, M., & Gurion, B. (1999). As 'process' as it can get: Students' understanding of biological processes. *International Journal of Science Education, 21*(12), 1281–1292.

Campbell, N. A., Reece, J. B., Taylor, M. R., & Simon, E. J. (2008). *Biology: Concepts & connections* (5th ed.). Benjamin Cummings.

Chi, M. T. H., Slotta, J. D., & De Leeuw, N. (1994). From things to processes: A theory of conceptual change for learning science concepts. *Learning and Instruction, 4*(1), 27–43.

Gilbertsson, M., Lind, A.-K., & Tersmeden, M. (2019). *Värmekamerans användning inom jordbruket.* https://docplayer.se/154824305-Varmekamerans-anvandning-inom-jordbruket.html. Accessed 4 Feb 2020.

Haglund, J., & Henriksson, F. (2021). Introducing infrared cameras in the study of pigs' physiology and health as cognitive apprenticeship in vocational education. *Action Research and Innovation in Science Education, 4*(1), 5–11.

Haglund, J., & Schönborn, K. J. (2019). The pedagogical potential of infrared cameras in biology education. *The American Biology Teacher, 81*(7), 520–523.

Haglund, J., Jeppsson, F., Hedberg, D., & Schönborn, K. J. (2015). Students' framing of laboratory exercises using infrared cameras. *Physical Review Special Topics – Physics Education Research, 11*(2), 020127.

Hmelo-Silver, C. E., & Pfeffer, M. G. (2004). Comparing expert and novice understanding of a complex system from the perspective of structures, behaviors, and functions. *Cognitive Science, 28*(1), 127–138.

Jones, H. G. (2004). Application of thermal imaging and infrared sensing in plant physiology and ecophysiology. *Advances in Botanical Research, 41*, 107–163.

Lahiri, B. B., Bagavathiappan, S., Jayakumar, T., & Philip, J. (2012). Medical applications of infrared thermography: A review. *Infrared Physics & Technology, 55*(4), 221–235.

Leguillon, P. (2019). *Using infrared cameras in education*. Karlstad University.

Pedaste, M., Mäeots, M., Siiman, L. A., de Jong, T., van Riesen, S. A. N., Kamp, E. T., et al. (2015). Phases of inquiry-based learning: Definitions and the inquiry cycle. *Educational Research Review, 14*(February), 47–61.

Planinšič, G., & Vollmer, M. (2008). The surface-to-volume ratio in thermal physics: From cheese cube physics to animal metabolism. *European Journal of Physics, 29*(2), 369–384.

Simpson, W. D., & Marek, E. A. (1988). Understandings and misconceptions of biology concepts held by students attending small high schools and students attending large high schools. *Journal of Research in Science Teaching, 25*(5), 361–374.

Soroko, M., & Howell, K. (2018). Infrared thermography: Current applications in equine medicine. *Journal of Equine Veterinary Science, 60*(January), 90–96.

Unger, S. D., Rollins, M. A., & Thompson, C. M. (2020). Hot-or cold-blooded? A laboratory activity that uses accessible technology to investigate thermoregulation in animals. *The American Biology Teacher, 82*(4), 227–233.

Wong, C. P., & Subramaniam, R. (2020). Use of technology in biology education – Case of infrared thermal imaging. *Journal of Biological Education*, 1–13. https://doi.org/10.1080/00219266.2020.1808510

Xie, C., & Hazzard, E. (2011). Infrared imaging for inquiry-based learning. *The Physics Teacher, 49*(6), 368–372.

Part III
Using Infrared Cameras in Science Teaching Practice

Chapter 9
Infrared Cameras as Smartphone Accessory: Qualitative Visualization or Quantitative Measurement?

Michael Vollmer

9.1 Introduction

In the past 30 years infrared cameras have become versatile tools in science as well as industry, demonstrated by an ever growing number of applications (Vollmer & Möllmann, 2018a). Progress in microsystems IR detector technologies has not only led to cameras with improved detector specifications, but also to focal plane arrays with above one million pixels. Whereas originally, IR cameras could be afforded by scientists and well educated engineers only, the price drop with mass production has also opened the field for non-specialists.

Nowadays, a wide variety of IR cameras is available, with expensive high end research cameras for scientists (some in the range above 100,000 €), medium priced cameras for engineering and industrial applications (10,000 € to 50,000 €) but also rather inexpensive cameras in the range of 1000 € or even below. The latter are attractive for craftsmen, but particularly also for science or physics teachers as remote contactless 2-dimensional temperature measurement makes IR cameras excellent visualization tools in teaching physics.

The potential of infrared imaging for educational purposes was already proposed 20 years ago (Karstädt et al., 2001). At that time, costs of cameras were well beyond regular school budgets. Meanwhile, dozens of respective demonstration and laboratory exercise experiments have been reported (Haglund et al., 2015, 2016; Kácovský, 2019a, b; Kubsch et al., 2017; Melander et al., 2016; Oss, 2019, 2020; Sieg et al., 2019; van Doremalen et al., 2019; Vollmer & Möllmann, 2018a, b, Wong & Subramaniam, 2018; Xie & Hazzard, 2011; Xu et al., 2019). Besides standard IR cameras (nowadays standard at least 640 x 480 pixels), many recent teaching

M. Vollmer (✉)
Department of Engineering, University of Applied Sciences Brandenburg, Brandenburg an der Havel, Germany
e-mail: michael.vollmer@th-brandenburg.de

J. Haglund et al. (eds.), *Thermal Cameras in Science Education*, Innovations in Science Education and Technology 26, https://doi.org/10.1007/978-3-030-85288-7_9

studies used cameras in the price range below 1500 € which had only limited spatial resolution. The latter has enormous influence for quantitative interpretation (see below), therefore, for convenience, the IR pixel numbers of the used cameras are provided at the end of each reference source in the reference list.

Initially, the inexpensive cameras were stand-alone products and not typical consumer products. This changed when inventing IR cameras which are either already included within a smartphone (Caterpillar Inc. (2020) or – more common – which can be used as smartphone accessory (FLIR Systems Inc., 2020; Seek Thermal Inc., 2020). The smartphone just acts as IR camera display and provides the operating software. These systems are mass products with prices of a few hundred € only. Therefore, widespread often playful use by laypersons, semiprofessional use by craftsmen as well as use in physics and science teaching at school is expected to dramatically increase within the next few years. As a matter of fact, such IR cameras will probably become standard school equipment for teaching at all levels.

Although such use is highly desirable, the introduction of IR imaging in schools must be accompanied by training offers unless one is already satisfied by colorful images as qualitative visualization. The physics of IR imaging is a complex field and quantitative interpretation of the recorded IR images requires detailed knowledge of the involved physics as well as of how the camera and its software operate and what a user must provide as input data to the camera software (Vollmer & Möllmann, 2018a).

The present chapter shall provide some respective background information. After presenting a comparative overview of currently available smartphone based IR cameras, it discusses some basic features of all inexpensive IR cameras, which are relevant for quantitative analysis. Focus will be on the smartphone types, but the arguments also apply to all other inexpensive IR cameras of say below 1500 €. Features include several parameter settings, time resolution, temperature accuracy and resolution as well as spatial resolution. In addition, software enhancements are discussed which produce nice images, but need to be considered with extreme care, as they also introduce artefacts due to image processing.

The discussion will include several examples, all recorded with a smartphone based IR camera of 160 × 120 pixels. As a matter of fact, practically all already demonstrated and published physics and chemistry experiments (e.g. Haglund et al., 2015, 2016; Kácovský, 2019a, b; Kubsch et al., 2017; Melander et al., 2016; Oss, 2019, 2020; Sieg et al., 2019; van Doremalen et al. 2019; Vollmer & Möllmann, 2018a, b, Wong & Subramaniam, 2018; Xie & Hazzard, 2011; Xu et al., 2019) may also be performed with a smartphone IR camera. For some examples on frictional effects, optics of glass, wavers and opaque plastic or thermodynamics, see e.g. Vollmer and Möllmann (2018b, 2020). Here, focus will mostly be on some lab and out-of-school recordings, which will of course show the same features as images in the classroom (e.g. thermal reflections, limited spatial resolution, artefacts due to image processing).

9.2 Smartphone Based IR Cameras

At present, three different commercial IR cameras associated with smartphones are available, all operating in the longwave (LW) thermal IR wavelength range from around 7.5 μm to 14 μm with microbolometer detectors. On the one hand, IR camera modules can be attached to smartphones or tablets (iOS or Android system), a technique offered for the *FlirOne* (FLIR Systems Inc., 2020) and *SeekThermal* (Seek thermal Inc., 2020) cameras. On the other hand, *Caterpillar* (Caterpillar Inc., 2020) offers an integrated IR camera (FLIR Lepton core) within the smartphone. In general, the most important specifications for IR cameras concern spatial and time resolution as well as temperature resolution.

Table 9.1 compares the specifications of the three systems. They mostly differ in numbers of pixels of the sensor (which in the manufacturers' specifications is strangely described as thermal resolution, which might be misleading) and used infrared optics characterized by the field of view (FOV). Both together – with the distance to an object – determine the actual spatial resolution (object size per pixel) of a camera (see examples below). The thermal resolution varies appreciably between 80 × 60 pixels (Cat) via 160 × 120 (FlirOnePro) up to 320 × 240 Pixels (Seek thermal). Similar to visual cameras, the fields of view are determined by the optics of the IR lens. Assuming quadratic pixels, the vertical FOV can be easily estimated from the horizontal FOV and the pixel numbers. The various camera models have slightly different optics. The best quantity to describe the spatial resolution in terms of angle is the instantaneous field of view (IFOV), i.e. the field of view of a single pixel ($IFOV=FOV/N_{pixel}$ with N_{pixel} being the number of pixels of the focal plane array (FPA) in the given direction). For later quick estimates of how much of an object is imaged onto a single pixel it is useful to give the angle in rad. For example, if we assume an object in a distance of 1 m, an IFOV of 6 mrad/pixel means that radiation from an object area of 6 mm × 6 mm falls onto a (quadratic) single pixel. No structures of an object below this size can be distinguished. Observing more distant objects, say in 20 m distance, the respective object sizes will already be ≈12 cm. However, if not perfectly aligned, such object radiation will realistically irradiate at least two pixels simultaneously, but none of them fully. Therefore, as a rule of thumb, object radiation must at least irradiate three neighboring pixels in order to be able to measure the correct object temperature from the central pixel (Vollmer and Möllmann 2018a; see also Fig. 9.1c below).

Recommendation Quantitative analysis requires that the spatial resolution must always be individually estimated for object scenes.

Frame rates are mostly limited below 9 Hz (more expensive camera systems do not suffer from this restriction!) which leads to somewhat awkward looking thermal videos. The frame rates of any camera are related to the detector integration times for an individual image. Microbolometer detectors have response times of around 10 ms. For accurate measurements, an object must be detected by a pixel for at least

Table 9.1 Comparison of properties of some available IR cameras within/for smartphones. If several models are available, the data refer to the model with best thermal resolution (status quo 08/2020)

Camera Property	Caterpillar S60/S61, 2 models	FlirOne, 3 models	SeekThermal, 3 models
Price in €	≤500/700–900	≤450	≤450
Thermal resolution (pixels)	80 × 60	both 80 × 60 and 160 × 120 (Pro)	both 206 × 156 and 320 × 240
Integrated VIS camera	Yes	Yes: 1440 × 1080	No
Horizontal field of view HFOV	46°	50° or 55°	36°, 20°, 32°
Best camera model instantaneous field of view = IFOV	0.575°/pixel = 10 mrad/pixel	0.344°/pixel (55°) = 6.0 mrad/pixel	0.1°/pixel = 1.75 mrad/pixel
Frame rate	8.7 Hz	8.7 Hz	< 9 Hz, <9 Hz, >15 Hz
T range in K	−20 °C to 120 °C (400 °C)	−20 °C to 400 °C	−40 °C to 330 °C
T accuracy	± 5 °C or ± 5% (the larger value)	± 3 °C or ± 5% (the larger value)	Not given
T sensitivity	150 mK	70 mK	<70 mK
Emissivity settings	4 preset values: 0.95; 0.80; 0.60; 0.30	4 preset values: 0.95; 0.80; 0.60; 0.30	Only pro allows four preset values 0.97; 0.8; 0.6; 0.3
Main advantages	No extra equipment, within smartphone, includes laser distance meter, Flir software	Simultaneous VIS image, Flir software: IR image processing + VIS image overlay	Highest thermal resolution
Main drawback	Low resolution	Undocumented image processing	Post recording analysis software missing

three time constants, i.e. around 30 ms integration time, which automatically limits useful frame rates to at maximum 30 Hz. This is also the typical frame rate of more expensive microbolometer cameras. For the smartphone cameras, the restriction to lower values is artificially introduced and not related to response time of detectors. As a result, they suffer from the same problem as all microbolometer cameras, i.e. a very limited time resolution. It manifests itself in cometary like blurred features in an IR image if objects move already at moderately small velocities of around 1 m/s (Vollmer & Möllmann, 2018a) with quite large T- errors, e.g. $\Delta T > 20$ K for an object of 340 K (Vollmer & Möllmann, 2018a; Chap. 2).

Recommendation Quantitative analysis of temperatures should mostly be restricted to still images.

Temperature ranges are comparable for the FlirOne and the SeekThermal whereas the Cat can only be used up to 120 °C, which for teaching is, however, still acceptable. The differences in temperature accuracy and sensitivity are less pronounced. The absolute accuracy is lowest for The FlirOnePro with ±3 °C or ± 5% of °C readings (the larger value must be applied). For the Cat with another Flir camera core, this is similar (though slightly larger) whereas no documentation could be found in specs or manual from the SeekThermal manufacturer. For high temperature readings, such absolute uncertainties can become unexpectedly large and may be neglected by unexperienced casual users of such cameras. For example, if measuring temperatures above 60 °C, the percentage error will apply. Temperatures of 100 °C or at the range limit of 400 °C would have accuracies of ±5 °C and even ±20 °C. More often, one is interested in relative temperature changes. Here, the temperature sensitivity defines the quality. It is below 70 mK for the SeekThermal and FlirOnePro, but about a factor of 2 larger for the Cat. These values are sufficient for most teaching purposes. We note, that more expensive cameras have much lower values.

Recommendation Temperature related specifications of smartphone based IR cameras are not great, but – if all other conditions are fulfilled – sufficient for teaching purposes.

Besides the camera related restrictions for quantitative analysis, the single most important source of potential error for analysis is always the user. Quantitative information in IR images depends on quite a few user input data in the operating and analysis software, describing the measurement conditions – and if the input data are erroneous, so too will be the results. This unfortunately very often leads to nice colorful images but wrong interpretations. Normally in IR imaging, the user must provide at least the following data: first, distance to object (which is usually known or easy to guess or measure), second ambient and surrounding temperatures (which result in thermal reflections), third relative humidity (which in combination with distance describes attenuation of object radiation) and most importantly emissivity ε. This is the biggest problem as each object will have its own specific value which – due to changes of surface roughness etc. - may easily differ from tabulated values of the same or similar materials. For analysis, ε must always be estimated or independently measured.

A serious disadvantage of smartphone based IR cameras is that they usually do not allow to use distance from camera to object, ambient temperature, and relative humidity as input parameters. The more advanced models do at least allow to use one of four given preset emissivity values. In comparison, quantitative industrial or research cameras allow a variation of ε with increments of 0.01. To judge potential errors due to wrong emissivity, one can analyze e.g. a blackbody source ($\varepsilon = 0.98$) with temperature of 100 °C at distance 1 m, $T_{\text{ambient}} = 20$ °C, and 50% relative humidity. For correct $\varepsilon = 0.98$, the camera shows $T = 99.6$ °C close to expectation. Allowed preset values of 0.95, 0.88, or 0.80 would give temperature readings of 101.2 °C, 106.1 or 112.4 °C. This example demonstrates that one may easily measure wrong temperatures by more than 10 °C if the wrong preset value

ε is selected. Furthermore, the correct value may be between two selectable values which automatically will lead to a systematic error.

Attention Not all needed data can be used as input for smartphone based IR cameras. Resulting errors can become quite large.

In addition to these considerations, analysis of IR images of any camera poses many more problems. For example, any object scene will usually be characterized by a spatial distribution of emissivity, i.e. any IR image which is displayed for a constant given value of emissivity is in fact a combination of temperature and emissivity contrast. And just attributing observed signal differences to temperature alone introduces errors (Vollmer & Möllmann, 2018a; Chap. 2). In addition, if there are nearby (not imaged) objects of different temperature, their thermal reflections may show up in the IR image. If not properly attributed, it may erroneously suggest an object property. Outdoor images suffer from additional complications. They depend, e.g., on wind speed, solar irradiation, view factor effects, shadows, night sky radiant cooling etc., which further complicate any quantitative analysis. Finally, analysis software for the CAT and FlirOne cameras already includes image processing (for nicer looking images) before the data are displayed. This can unfortunately lead to severe errors as shown below. Post recording analysis is in principle possible for the FLIR systems, however, the cost of really versatile software (usually developed for more expansive cameras) can easily be above the cost of the camera itself. In comparison, the SeekThermal cameras unfortunately offer no post recording analysis software at all.

Recommendation Every IR image needs a very careful individual analysis, taking into account all potential effects. As discussed with examples below, my personal advice is to use all smartphone based IR cameras for qualitative use which is well suitable for science and physics teaching in school. Only in a few special well defined cases in the lab, quantitative analysis is possible and useful, then however requiring a much deeper user knowledge.

Each of the three systems has specific advantages. The fastest set up time is obviously possible for the integrated camera of the *CatS60/61* (no time for connection and start up needed). In addition, as the only camera in the list, the *CatS61* has the advantage of an integrated laser range meter for measuring object distances. The FLIROne comes with a visual camera, i.e. each IR image is automatically also recorded in the visible (VIS) range. The latter information is used for image processing which has some advantages for qualitative display (though definitely not for quantitative analysis), e.g. using image overlay features. The SeekThermal software seems to also offer image overlay, though this is poorly documented. However, the SeekThermal has the largest thermal spatial resolution. The typical battery life time (FlirOne, Seek) during operation depends on use, but is definitely below 1 h. Therefore it is recommended to have a power bank available for charging during operation. The main drawbacks are also easily summarized. The Cat has the lowest spatial resolution. The (unexplained) image processing software for The FlirOne and Cat introduces artefacts (see below) and the SeekThermal does not offer post recording analysis tools at all.

The post recording analysis, allowing e.g. line plots, may be the reason why the FLIR system cameras are more popular. Therefore all of the following image examples were recorded with a FLIROnePro camera (the Cat uses similar software and behaves alike). All cameras have the choice of several false color palettes. In order to avoid confusion, the same one was used for all examples. For more detailed information on the choice of color palettes for IR images with regard to embedded information, see (Vollmer & Möllmann, 2018a).

Since the camera development is very fast, new camera models with higher thermal resolution may become available soon. This seems also possible for the Cat smartphones which use the 80 × 60 pixels FLIR Lepton cores. Meanwhile, new Lepton 3.5 cores with 160 × 120 pixels are available which – in new Cat models – may also extend the resolution to 160 × 120 pixels. Hopefully, all software related problems will be fixed or at least explained in the future.

9.3 Examples of Spatial Resolution of Smartphone Based IR Cameras

For most practical cases in physics teaching, the spatial resolution will be a dominant problem. Its influence on measurements is usually analyzed by placing a slit of given width in front of a black body source and measure line plots of temperature across the slit. Figure 9.1 shows a typical result for a more expensive 320 × 240 commercial camera (after Vollmer & Möllmann, 2018a). The measured signal behind a slit of given width decreases with decreasing slit width. Plotting the result versus viewing angle rather than slit width, it becomes obvious, that correct temperature readings are only expected if the slit angle is above around 3 IFOV. This supports the qualitative argument from above.

Repeating the experiment for the FlirOnePro (different distance, different IFOV) and using either the free app FlirTools or other professional software for the analysis, the results are similar, but differ in two aspects. Again, correct temperature readings may be expected at the center if object sizes corresponding to at least 3 IFOV are imaged. The first strange fact, however, is that although IR images are based on radiometric jpg images with 160 × 120 pixels, they are stored as 640 × 480 pixel images. Of course IR spatial resolution is not changed, but the displayed images suggest better results. As a result, the line plot in Fig. 9.2b (below) which should refer to only 30 IR pixels corresponds to around 120 pixels, i.e. it is a factor of four larger than expected.

The used interpolation process is not publicly available. Unfortunately, the raw data image processing produces unwanted artefacts for quantitative analysis

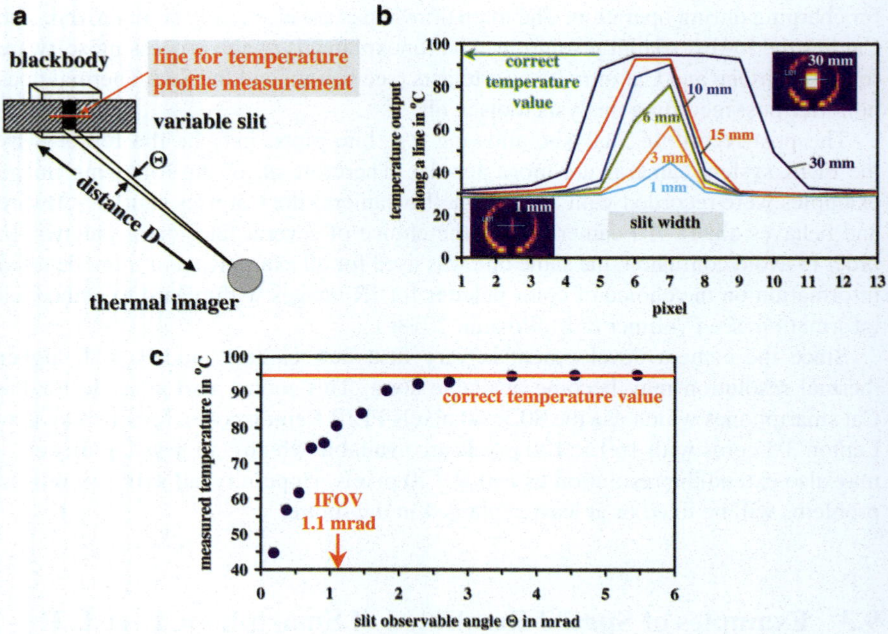

Fig. 9.1 Spatial resolution of an IR camera. (**a**) set up for measurement (here, D = 5.5 m), (**b**) T-profiles across the slit, (**c**) Temperature versus observation angle. One needs at least 3IFOV (in this case 18 mm) for correct temperature reading

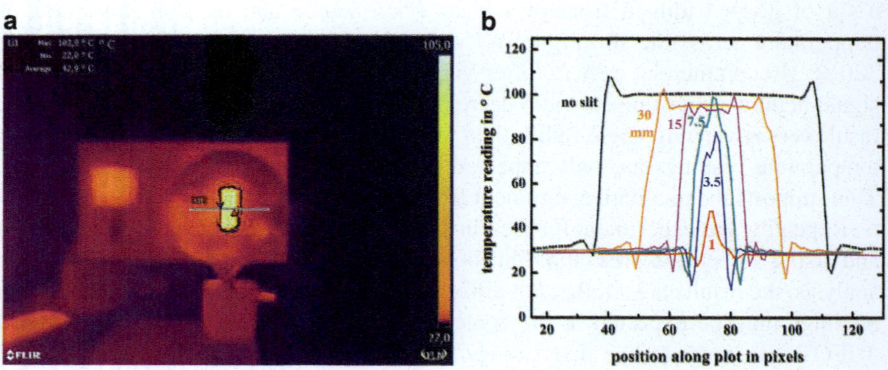

Fig. 9.2 Investigation of the spatial resolution of a FlirOnePro camera (for details, see text)

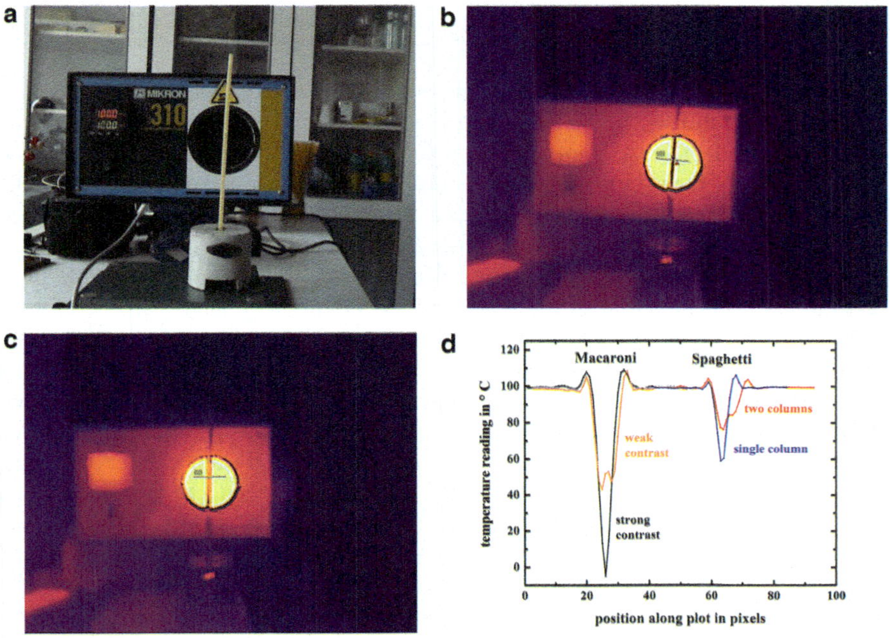

Fig. 9.3 Macaroni and spaghetti in front of a black body. (**a**) set up, (**b**) tilted pasta, (**c**) filling two columns, (**d**) analysis

(although it may be helpful for qualitative visualization). Figure 9.2 shows that near edges, the temperature values are never correct. In the present case, these artefacts led to T differences of around 10 K, way above any specified temperature accuracy. Worse even, reducing the slit width means that the overshooting features (which are quite characteristic for any Fourier Transform mechanism) come closer together. At a critical distance (here happening between 15 and 7.5 mm) they start to overlap. As a consequence, they give much higher T values at the center than could be expected.

Another complementary example is shown in Fig. 9.3, where slits – transmitting radiation – were replaced by thin obstacles, weakening the radiation. We used room temperature raw pasta, macaroni and spaghetti, placed in front of the black body. With proper distance and horizontal adjustment, it was possible that the spaghetti was imaged either on one detector column only, or on two, a purely geometrical effect with object size corresponding to about one IFOV. As part of the pixels still received radiation from the adjacent blackbody, the spaghetti temperature was always higher than expected. However, changing between a perfect vertically aligned thicker macaroni to a slightly tilted one led to stronger changes in the processed images with much more overshooting. Now, surprisingly, temperatures of the macaroni even below freezing temperature was measured, obviously a total nonsense result!

9.4 Summary of Artefacts

Due to image processing, the apparent overshooting and undershooting signals are way off from realistic values and it is no longer possible to estimate accurate temperatures of objects. Even without image processing, one needs at least three times the IFOV for accurate measurements of the central pixel, i.e. 18 mrad. This would refer to an object of 18 mm size at a distance of 1 m, 18 cm size at a distance of 10 m, or ≈3 mm at the minimum focusing distance of 0.15 m! Taking the artefacts due to the image processing also into account will require even larger objects, resembling a much lower spatial resolution for quantitative analysis. Therefore it does not make much sense to observe a defect high voltage line or a small thermal bridge in a building insulation from a large distance with such a camera. Similarly, a thin current carrying wire of 0.5 mm diameter which heats up due to the dissipated electrical energy will probably just show up a little bit if observed from a distance larger than 0.5 m, but have a much better contrast if observed from smaller distances. Respective demonstration experiments are easy and allow students to characterize the spatial resolution and limitations of such cameras.

We end the discussion of spatial resolution by presenting two practical outdoor observations. Figure 9.4 depicts an airplane approaching San Diego airport briefly before touchdown. Not knowing accurate dimensions one may still do rough estimates. Medium range airplanes may have total lengths of about 40 m. Compared to the total field of view of 55° (160 pixels), the length corresponds to about 15°, i.e. 0.26 rad (43 pixels). The thermal spatial resolution is therefore around 1 m/pixel. Assuming for simplicity that the image was recorded more or less vertically straight up, the height of the plane in the image is around 150 m. One can easily see the hot exhaust gases from the engine. Engine diameters are around 2 m or larger, which would correspond to maybe 2 IR pixels, only. The image processing has increased this number to around 8 manipulated pixels. The small thin structures at the end of the wings are much smaller and do not show up at all.

Fig. 9.4 Descending airplane near San Diego airport (for details, see text)

Fig. 9.5 Golden Gate Bridge, observed from Battery Spencer. The varying distance to structural elements nicely illustrates the varying spatial resolution in the image (for details, see text)

Figure 9.5 depicts the Golden Gate Bridge in San Francisco from the probably nicest northwestern view point at Battery Spencer. The top main cable has a diameter of 0.92 m, each vertical suspender rope (in bundles of 4) around 7 cm. Distance from the camera to the closer main tower (base dimension 10 m $\times$ 16 m) is around 300 m, to the middle of the bridge around 900 m and to the distant main tower 1500 m. In these distances, and using an IFOV of 6 mrad, we find around 1.8 m/pixel, 5.4 m/pixel, and 9.0 m/pixel. Indeed, the close tower has a width of around 5 to 6 thermal pixels.

The main cable close to the main tower covers only around half a thermal pixel, however, as it is heated up by absorption of solar radiation, its contribution is much larger than the sky contribution to the same pixel and it shows good contrast to the colder background sky including clouds. Quantitative analysis, however, makes no sense at all. One may still guess the suspender ropes close to the main tower, although each only accounts for about $0.07/1.8 \approx 4\%$ of the pixel area. However, they are also heated up and contribute a large signal whereas the 96% of sky background of lower temperature irradiating the pixels contribute a disproportionally small signal, only.

The different ratio of object and sky background signals on a pixel also explain why the main tower appears much brighter (warmer) than the main cable, which itself appears warmer than the suspension ropes. The main cable contribution to the signals decrease with increasing distance, as do the signals of the suspension ropes. They are only seen for good contrast to cold clear sky and disappear for the warmer cloud background and even the water of the bay. The latter is usually quite cold for swimmers, but of course warmer than clouds and sky. Also, they are hardly seen close to the center of the bridge.

Last but not least, one can easily see a thermal reflection of the warm bridge on the cold water due to the contrast with the adjacent reflected colder sky background. In contrast, in the VIS image, the respective shadow part is darker.

Recommendation If only discussing spatial resolution, it is clear that quantitative analysis with IR cameras is not as easy as it seems. We strongly recommend only qualitative use in school demonstrations to avoid errors and misinterpretations. Even those are sometimes complex as the examples of the bridge have demonstrated, i.e. each IR image always requires an individual discussion.

9.5 Image Processing: Boon or Bane

The Cat and Flir cameras have another software feature, which for some cases is helpful but which can also lead to potential misinterpretations, in particular of untrained users. As simultaneous VIS images are recorded, the software uses the higher resolution VIS image and applies edge recognition algorithms. The parallax between the two images can be manually corrected (the lenses of the VIS and IR cameras are about 1 cm apart). The edges from the VIS image can then be superimposed onto the IR image. This may help as objects are easier identified and people using such cameras as toys may like the feature. However, when teaching physics or science, it must always be kept in mind, that these lines are visible light artefacts and have nothing to do with the IR image. Figure 9.6 shows an example of an alarm-clock and a smartphone in the visible range, in pure IR and as so-called MSX, i.e. the IR images with superimposed information from edge recognition algorithms. For any kind of measurement or qualitative visualization

Fig. 9.6 Demonstration of overlay of edge features from a visible photo onto the IR images using the so called MSX technique for the Flir and Cat cameras

of thermal effects in teaching, the superimposed information is – to my personal opinion – distracting. It again just suggests higher spatial resolution in the IR image than actually present but does not add any other information related to IR radiation. Unexperienced users may just intuitively, but wrongly interpret the additional information as relevant for the thermal object features.

9.6 Everyday Use: Aesthetic Infrared Views

The main and only reason why IR smartphone based cameras are available is that manufacturers can make profit. The target group is not just physics and science teachers or in general people interested in natural sciences having found ideal X-mas or birthday presents, but everybody interested in technical gadgets which can be used with a smartphone. Therefore, besides physics and science topics, the majority of users just want to have fun. They will probably record many nice and colorful images of everyday life, from holidays, in the wilderness etc. In this area, qualitative visualization dominates and the aesthetic views of seeing in the infrared may perhaps lead to a new group of semiprofessional infrared photographers (who very likely will soon look for higher spatial resolution systems).

In the following, two examples are presented where the IR images (in the opinion of the author) have much more appeal than the visual ones and may come close to photographic art. Figure 9.7 shows a view from Millennium Bridge. The IR image Fig. 9.7a (after Vollmer & Möllmann, 2020) offers a spectacular aesthetic, even

Fig. 9.7 A hot summer day in London near St. Paul's cathedral

artistic, view of St. Paul's Cathedral in London, probably more so than recordable with any visible light camera.

Besides, it may also be discussed with regard to some physics. The black color traffic light at the lower left has heated up appreciably and clearly stands out as thermal feature. The green signal for pedestrians (probably with LEDs) is of course not detectable at all in the IR due to its relatively tiny thermal input compared to the solar absorption. True, the temperature range – chosen for the overall appearance of the image – was huge from −20 °C to 40 °C at $\varepsilon = 0.9$, however, the signal did also not show up in post-recording analysis with a range below 2 K. A second interesting feature relates to the spatial resolution. Without knowing the VIS image, it would be extremely difficult to detect any pedestrians, not even the closest ones. Measuring the size of the closest single person within the VIS image while assuming a typical height of 1.8 m and the known vertical FOV of 55°, the 1.8 m correspond to around 0.1 rad. This yields around 11 cm/IR pixel. Being slightly tilted, the person may have a horizontal width of 30 cm, which would correspond to at most three IR pixels. From the Golden Gate bridge example (above) one would therefore expect the person to show up in the IR image if the combined thermal and emissivity contrast to the surrounding is large enough. Unfortunately, it was a hot day and people's clothes are heated up similar to the background pedestrian zone and there is not much of an emissivity difference.

Conclusion Even if spatial resolution is in principle sufficient – objects may still be veiled if thermal contrast is low.

Finally, Fig. 9.8a depicts another spectacularly looking false color IR image, a jungle temple scene at Angkor Wat in Cambodia. The huge brightness differences in the VIS image between sky and dark foreground are less pronounced in the IR. Due to the angle of incidence, roofs had received more solar radiation and show up warmer than the walls. Again, the people in front (distance of around 27 m) are hardly detectable in the IR and only show up prominent in the VIS image due to the white clothes. Their width corresponds to at most 10 mrad, i.e. just 1.5 IR pixels. Similar to the above example, thermal contrast is low, so they are nicely veiled in the IR image. Thermal contrast is, however, different for the tree branches with colder sky background (see also bridge example above). This nicely allows one to visualize the spatial resolution effect. The tree top with smallest branches is in a distance of around 40 m. The smallest branches in the upper center of the image are hardly detectable. They were estimated to cover around 2 mrad (thickness 8 cm) and 1 mrad (4 cm) only, i.e. well below the 6 mrad/pixel IFOV. Only due to the much larger thermal contrast to the sky are they still just detectable similar to the suspension ropes of the Golden Gate bridge example.

There is hardly a limit of what can be imaged with IR cameras, which also holds for the smartphone based systems. The opportunity of fast set up will e.g. also allow for snapshots of animals, if they are not moving. Besides others, cats and dogs, racoons, sea gulls, seals, or monkeys have been captured and even fast moving bats show up as shadowlike features in IR movies. Other popular items for qualitative imaging in nature are the sun (caution, avoid looking directly into it with an IR camera in order to prevent damage to the sensors) or the moon (Shaw

Fig. 9.8 Jungle temple scene at Angkor Wat/Cambodia

et al., 2015; Vollmer & Möllmann, 2018a) although the latter with 0.5° angular diameter will–even for full moon–only correspond to 1.5 pixels for a 160×120 FLIR OnePro camera. Other fascinating natural thermal features are geysers or hot pools. Again quantitative analysis can become difficult, and not only due to limited spatial resolution (Shaw et al., 2017).

9.7 Conclusions

Smartphone based IR cameras are fascinating technical gadgets which may prove very helpful in science and physics teaching at all school levels. However, being low end mass products, they have quite a few limitations, in particular regarding spatial resolution, variation of input parameters, image processing and post-recording analysis (which is possible, but may be expensive). Therefore, whenever a quantitative analysis is attempted, extremely careful discussion of each individual image is needed. The required background knowledge of IR imaging with all its problems is typically beyond what a physics teacher knows. However, qualitative use is very attractive. There are well above a hundred suitable published experiments which can be successfully used in teaching (see Vollmer 2018a), in particular whenever temperature differences are visualized and objects are close, e.g. in lab experiments, such that spatial resolution is sufficient. It is therefore highly recommended, that for the majority of teaching applications, these cameras should be used qualitatively. Only in a few special well defined cases in the lab, quantitative analysis is possible and useful, then, however, requiring a much deeper user knowledge.

Acknowledgments I want to thank Klaus Peter Möllmann for our long collaboration and inspiring investigations in the field.

References[1]

Caterpillar, Inc. (2020). *Introducing the new Cat® S62 Pro*. https://www.catphones.com/en_gb/cat-s60-smartphone.html. Accessed 19 Aug 2020.

FLIR Systems, Inc. (2020). *FLIR ONE Pro*. https://www.flir.com/products/flir-one-pro/. Accessed 19 Aug 2020.

Haglund, J., Jeppsson, F., Hedberg, D., & Schönborn, K. J. (2015). Thermal cameras in school laboratory activities. *Physics Education, 50*(4), 424–430. https://doi.org/10.1088/0031-9120/50/4/424. (60 × 60/80 × 60).

Haglund, J., Jeppsson, F., Melander, E., Pendrill, A. M., Xie, C., & Schönborn, K. (2016). Infrared cameras in science education. *Infrared Physics and Technology, 75*, 150–152. https://doi.org/10.1016/j.infrared.2015.12.009. (no information).

Kácovský, P. (2019a). The visit of a thermal imaging world in one physics lesson. *Physics Education, 54*(4), 045011. https://doi.org/10.1088/1361-6552/ab15b9. (140 × 140).

Kácovský, P. (2019b). Electric circuits as seen by thermal imaging cameras. *The Physics Teacher, 57*(9), 597–599. https://doi.org/10.1119/1.5135785. (140 × 140).

Karstädt, D., Möllmann, K.-P., Pinno, F., & Vollmer, M. (2001). There is more to see than eyes can detect: Visualization of energy transfer processes and the laws of radiation for physics education. *The Physics Teacher, 39*(6), 371–376. https://doi.org/10.1119/1.1407135. (320 × 240).

Kubsch, M., Nordine, J., & Hadinek, D. (2017). Using smartphone thermal cameras to engage students' misconceptions about energy. *The Physics Teacher, 55*(8), 504–505. https://doi.org/10.1119/1.5008354. (206 × 156).

Melander, E., Haglund, J., Weiszflog, M., & Andersson, S. (2016). More than meets the eye–Infrared cameras in open-ended university thermodynamics labs. *The Physics Teacher, 54*(9), 528–531. https://doi.org/10.1119/1.4967889. (80 × 60/160 × 120).

Oss, S. (2019). Infrared imaging of a non-stationary thermal conductive process and observation of its Green's kernel. *European Journal of Physics, 41*(1), 015102. https://doi.org/10.1088/1361-6404/ab4680. (160 × 120).

Oss, S. (2020). Infrared imaging of the cooling fin equation. *European Journal of Physics, 41*(5), 055102. https://doi.org/10.1088/1361-6404/ab9891. (160 × 120).

Seek Thermal, Inc. (2020). *Powerful thermal imaging cameras designed for your smartphone*. https://www.thermal.com/compact-series.html. Accessed 19 Aug 2020.

Shaw, J. A., Nugent, P. W., & Vollmer, M. (2015). Infrared moon imaging for remote sensing of atmospheric smoke layers. *Applied Optics, 54*(4), B64–B75. https://doi.org/10.1364/AO.54.000B64. (640 × 512).

Shaw, J. A., Nugent, P. W., Harris, W., & Vollmer, M. (2017). Infrared yellowstone. *Optics and Photonics News, 28*(6), 36–43. https://doi.org/10.1364/OPN.28.6.000036. (max single image 640 × 512).

Sieg, P. G., Berner, W., Harnish, P. K., & Nelson, P. C. (2019). A demonstration of the infrared activity of carbon dioxide. *The Physics Teacher, 57*(4), 246–249. https://doi.org/10.1119/1.5095383. (320 × 240).

van Doremalen, R. F. M., van Netten, J. J., van Baal, J. G., Vollenbroek-Hutten, M. M. R., & van der Heijden, F. (2019). Validation of low-cost smartphone-based thermal camera for diabetic foot assessment. *Diabetes Research and Clinical Practice, 149*, 132–139. https://doi.org/10.1016/j.diabres.2019.01.032. (160 × 120 and larger).

Vollmer, M., & Möllmann, K.-P. (2018a). *Infrared Thermal imaging: Fundamentals, research and applications* (2nd ed.). Wiley. (multiple, up to 1 Megapixel).

[1] Pixel count of respective thermal sensor from source included in brackets.

Vollmer, M., & Möllmann, K.-P. (2018b). Infrared cameras as accessories to smartphones: Facts you need to know. *Physics Education, 53*(6), 065019. https://doi.org/10.1088/1361-6552/aae277. (320 × 240/640 × 512).

Vollmer, M., & Möllmann, K.-P. (2020). Unsichtbares sichtbar gemacht: Infrarotkameras für Smartphones. *Physik in Unserer Zeit, 51*(1), 29–35. https://doi.org/10.1002/piuz.201901551. (160 × 120).

Wong, C. P., & Subramaniam, R. (2018). Seeing in a different light—Using an infrared camera to teach heat transfer and optical phenomena. *Physics Education, 53*(3), 035007. https://doi.org/10.1088/1361-6552/aaa5ee. (320 × 240).

Xie, C., & Hazzard, E. (2011). Infrared imaging for inquiry-based learning. *The Physics Teacher, 49*(6), 368–372. https://doi.org/10.1119/1.3628268. (80 × 80).

Xu, X., Wu, M., & Wang, X. (2019). Smartphone visualization of thermal phenomena with thermal imaging accessories. *Journal of Chemical Education, 96*(11), 2545–2552. https://doi.org/10.1021/acs.jchemed.9b00131. (160 × 120).

Chapter 10
An Infrared Camera: Multiple Ways to Use a Modern Device in Introductory Physics Courses

Gorazd Planinšič, Urška Nered, and Eugenia Etkina

10.1 Introduction

The demands on the amount of information that our students both in high schools and in college need to learn in a physics course increases every day. Teachers struggle to include physics discoveries of the twentieth and twenty-first centuries while making sure that their students learn the foundations of classical physics. Two years ago, two authors of this chapter (GP and EE) proposed a framework that allows teachers to incorporate modern devices in introductory physics courses without increasing the content load (Planinsic & Etkina, 2019). We use the term "modern device" for any device invented after the middle of the twentieth century. In this chapter we will apply this framework to the infrared camera (IR-camera).

An IR-camera allows us to see the world through an "artificial eye" that is sensitive to the electromagnetic waves of wavelengths larger than those of visible light. Similar to the discovery of radio telescopes that allowed scientists to see the part of the universe previously hidden from human eyes, the IR-cameras allow our students to see the aspects of phenomena that are hidden from them otherwise. We can think of an IR-camera as an affordance to work with a different representation. Representations and images are crucial in construction of new knowledge (Zull, 2002).

G. Planinšič (✉) · U. Nered
Faculty for Mathematics and Physics, University of Ljubljana, Ljubljana, Slovenia
e-mail: gorazd.planinsic@fmf.uni-lj.si

E. Etkina
Graduate School of Education, Rutgers, the State university of New Jersey, New Brunswick, NJ, USA
e-mail: eugenia.etkina@gse.rutgers.edu

J. Haglund et al. (eds.), *Thermal Cameras in Science Education*, Innovations in Science Education and Technology 26, https://doi.org/10.1007/978-3-030-85288-7_10

In this chapter, we discuss the framework and then describe the activities with an IR-camera that we used within each component of the framework. We share the responses that the students provide (or might provide) as well as suggestions on how to guide them through the activities. We also provide technical advice for designing the experiments.

10.2 Framework for Using Modern Devices in Introductory Physics

As we mentioned in the Introduction, the structure of the chapter is based on the framework for using modern devices in introductory physics, which has three components (see Fig. 10.1):

1. using a modern device as a black box,
2. learning how the modern device works and
3. learning new physics using the knowledge of how the modern device works.

The relationship among the components of the framework in Fig. 10.1, shown as a spiral, represents the increase of conceptual difficulty among the components.

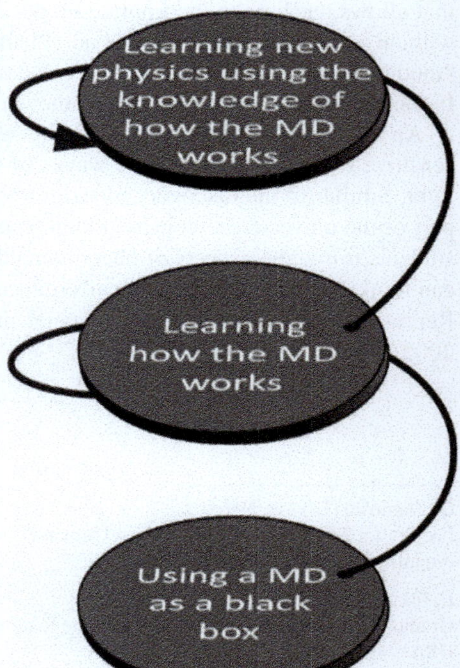

Fig. 10.1 A three-dimensional representation of the framework (Planinsic & Etkina, 2019)

After students learn concepts in one component, they are prepared for the next one. Using a spiral instead of a linear progression emphasizes that in the next component of the framework the students come back to the same modern device, but in a different context. We can think of the framework as being epistemological in nature rather than focusing on the technical details of the device's operation.

To implement activities described in this chapter successfully teachers need to be familiar with the fundamentals of thermal imaging. For the requisite information regarding the fundamentals of thermal imaging, see Chap. 2 of this book.

10.2.1 Using an IR-Camera as a Black Box

Using a modern device as a black box allows students to use an IR-camera to investigate phenomena without going into the physics mechanisms of the device itself. However, even though we do not seek explanations for how the device works, black boxes offer several opportunities for connecting, comparing and contrasting features of a modern device with other similar devices or with phenomena already familiar to the students.

Students can use an IR-camera as a black box for observing temperature changes of different objects while studying various physics topics. If the students have suitable equipment, they can perform the activities by themselves. Alternatively, the students can observe the IR-videos and/or images provided here and then continue with the activities described below. For teachers, it is important to choose carefully the materials from which the objects are made (for example, avoid using objects with low emissivity/high reflectivity unless this is purpose of the activity) and conditions in which the experiments are performed (for example, using gloves to minimize warming up the objects with hands).

10.2.1.1 Activity 1 – Pulling a Heavy Object Along a Rough Horizontal Surface

The following activity illustrates how an IR-camera can be used as a black box. The activity is suitable for the introductory class for the topic of work and energy (Etkina et al., 2020). Research shows that students have significant difficulties when trying to grapple with the conversion of mechanical into internal energy (Daane et al., 2015) as the latter is imperceptible, but when they have access to IR images, they can reason that mechanical energy is converted to internal thermal energy (Haglund et al., 2015). To be successful with the activity, the students need to be familiar with kinetic and potential energies and work as the means of changing the energy of a system when an object outside the system exerts a force on it and the point of application of the force moves a distance d.

Imagine that you are pulling a heavy box across a carpet. The box is moving at constant speed, therefore the sum of the forces exerted on it is zero. If the sum of

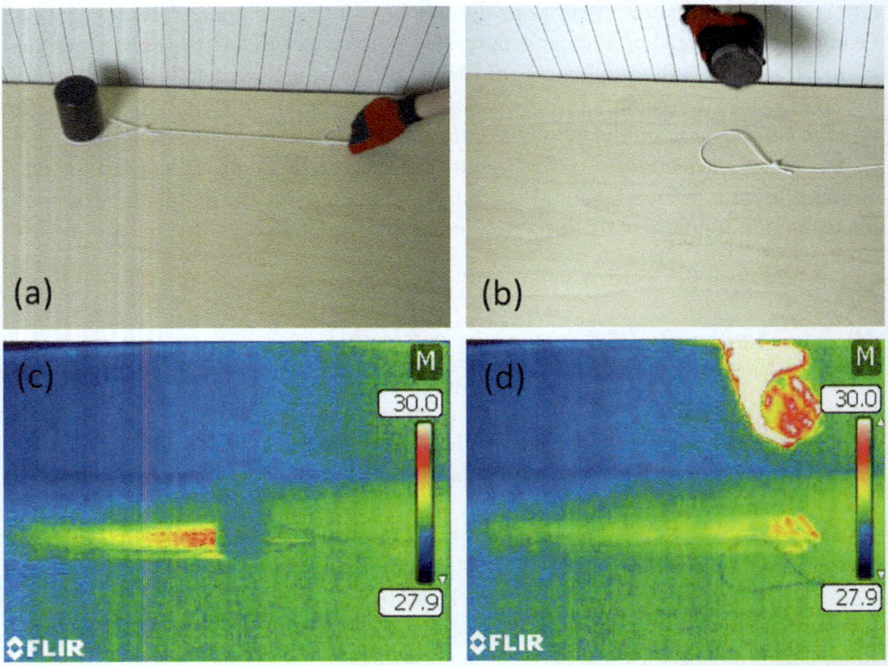

Fig. 10.2 Pulling a heavy metal cylinder across rough horizontal surface and then lifting the cylinder: (**a**), (**b**) photos taken with a regular (VIS) camera; (**c**), (**d**) photos taken with an IR-camera

the forces exerted on the box is zero, so should be the total work. Yet, when we examine the surface of the box and the carpet, we find them both slightly warmer. Where did this energy come from? Answering this question helps students construct the concept of the internal energy change.

Students watch the video of a person pulling a 5-kg metal cylinder across a wooden surface slowly, at constant speed https://youtu.be/QVWWytYQ0DQ. We used a 5 kg metal cylinder as the object and an unfinished plywood board as the surface. We glued a 3-mm thick foam board to the bottom of the cylinder and sand paper to the foam board. The role of the foam board was to reduce the transfer of thermal energy to the metal, therefore making the temperature changes of the sand paper and the table surface larger. To make the heavy object slide at constant speed across the rough surface without tipping over, we fixed the string that pulls the cylinder close to its bottom surface. The experimenter used a glove to minimize heating of the metal object by the hand. The video consists of two parts. The first part was obtained with a regular camera (Fig. 10.2a, b) and the second part is obtained with a FLIR B335 IR-camera (Fig. 10.2c, d).

We can pose the following questions to the students:

(a) Watch the video and describe how the surfaces of the board and the cylinder change as a result.
(b) Analyse the process. Choose the cylinder-board as the system and use your knowledge of work and energy. The person pulling the bob is outside the system. Remember, that internal forces do not do work on the system but allow to convert one form of energy into another.

The students notice that the temperature of the surface across which the cylinder was pulled and the bottom surface of the cylinder increased in temperature by about 1 °C. To explain the process using their knowledge of work and energy they might reason as follows:

> As the elevation of the cylinder did not change, Earth did not do work on the system. The only object that did work on the system is the string pulling the cylinder. The work is positive. Thus, the energy of the system should have increased, and yet, we cannot see the increase in speed, i.e. increase in kinetic energy. Where did the work go? As the students already noticed that the temperature of both surfaces increased, they might speculate that there is a new type of energy that increases because of the work done by the string. At this point, the teacher can help them with the name for this new type of energy –the internal thermal energy of both surfaces (Etkina et al., 2019, 2020).

10.2.1.2 Activity 2 – Shooting a Glass Marble Through a Cardboard

In Activity 1 the students reason qualitatively. The following activity allows them to reason quantitatively and to calculate the change in internal energy using the data from the infrared images. As the activity requires students to be familiar with thermodynamics (specifically the equation $\Delta U_{\text{int}} = mc\Delta T$; ΔU_{int} – change in internal energy of an object, m – mass of the object, c – specific heat of the object, ΔT – temperature change of the object), we suggest that they approach this activity while studying thermodynamics.

The process that students go through consists of first analysing the data for a glass marble shot from a slingshot into a cardboard (see Fig. 10.3).

The students can determine the speeds from the high-speed video (see video at https://youtu.be/nOBDWUg6vCc; the movie was recorded at 1000 fps using Sony RX10 IV camera). The next step is to explain qualitatively why the speed changed. The students usually explain that some of the initial kinetic energy of the marble was converted to internal (thermal) energy of the board and the marble. Applying the generalized work-energy theorem, students can describe this process mathematically: the initial kinetic energy is equal to the sum of the final kinetic energy and the change in internal energy: $K_{\text{i}} = K_{\text{f}} + \Delta U_{\text{int}}$.

Once the students discuss the conversion of kinetic energy of the marble into the thermal energy of the marble and cardboard, they examine the IR images of the cardboard before and after the marble penetrated it (see Fig. 10.4) and discuss whether the image is consistent with their explanation. The red spot on the second image indicates that the part of the cardboard where the marble penetrated it warmed up. Therefore, these data support their explanation qualitatively.

Using the IR images and suitable software, the students determine the average temperature of the ruptured and un-ruptured part of the cardboard (see Fig. 10.5a). The average temperature at the area where the marble penetrated the board was 23.3°C ± 2.2°C and at the other area 21.6°C ± 0.1°C. We made all IR images using a FLIR ONE Pro camera (see Chap. 9 in this book) in combination with an Android smartphone and analysed the IR images using FLIR Tools software.

In the next step, the students calculate the expected temperature change of the cardboard using the generalized work-energy theorem and compare the result with the temperature difference obtained from the IR image. They also indicate any assumptions that they made. At this point we give students additional data: average mass of the ruptured part of the cardboard (see Fig. 10.5b) $m_{cb} = 0.27$ g ± 0.02 g, which we determined through several experiments and the average specific heat of the cardboard $c_{cb} = 1400$ J/(kgK) ± 100 J/(kgK), which we determined based on data from various sources.

When calculating the change in temperature the students need to make the following simplifying assumptions (if they miss some, the teacher should help them) in addition to assuming that the temperatures obtained from IR images are the actual temperatures of the objects:

1. The change in gravitational potential energy of the system (marble + cardboard + Earth) during the collision is much smaller compared to the change in its kinetic energy and can be neglected. (The assumption can be validated using data above.)
2. The board is fixed and does not move, thus final kinetic energy of the board is negligible. (Students notice in the video, that the board only slightly moves after the marble penetrates it.)
3. The increase of the marble's internal energy is negligible compared to the increase of the internal energy of the cardboard (glass has low thermal conductivity).

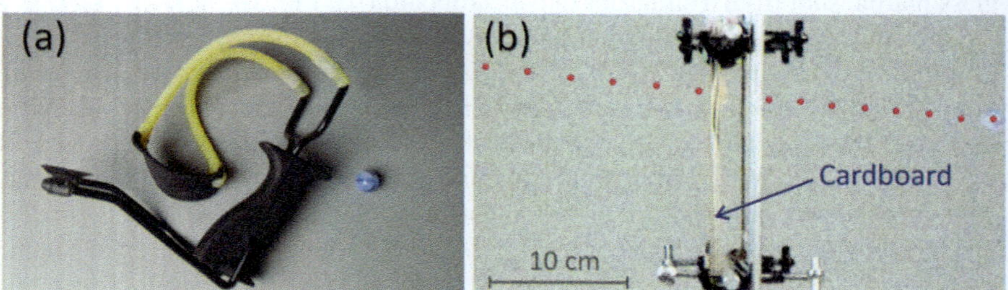

Fig. 10.3 (**a**) The slingshot and the glass marble, (**b**) Graphical analysis of the high-speed video (red dots show the positions of the marble in 1/1000 s time intervals). The data consist of the mass of the marble ($m_m = 8.33$ g) and its speed before and after passing through the cardboard ($v_i = 29.5$ m/s ± 0.4 m/s)

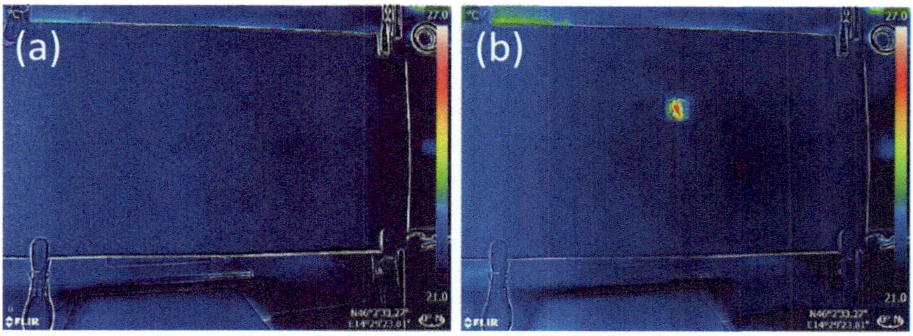

Fig. 10.4 IR images of the cardboard (**a**) before and (**b**) after the marble penetrated it. The contours (thin white curves) are obtained from visible light photos superimposed on IR images, to make it clearer where the edges of the cardboard are

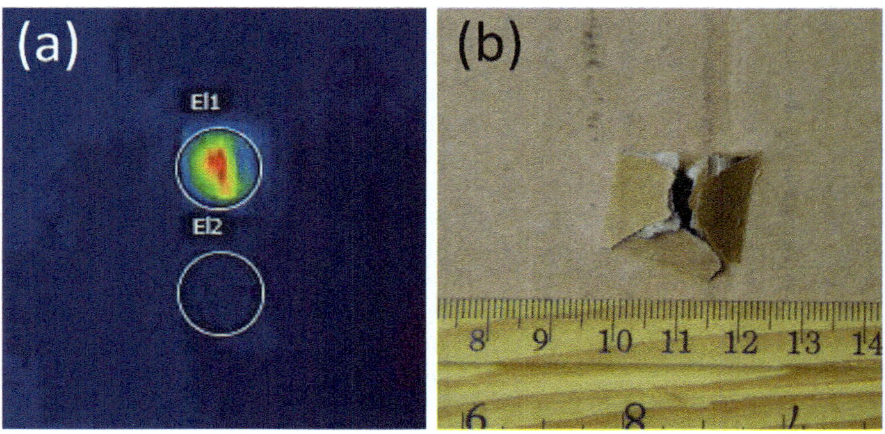

Fig. 10.5 (**a**) Part of the IR image (Fig. 10.4b) with marked areas where the average temperatures were determined. The colours on the images show the temperature in the interval from 21 °C (dark blue) to 27 °C (white). (**b**) Close view of the ruptured part of the cardboard

4. The initial temperatures of the marble and the cardboard are equal (we held the marble using thick gloves to ensure that the marble and the gloves are at room temperature before shooting).

Using the generalized work-energy theorem and given data, the expression for the increase of the temperature of the ruptured part of the cardboard is:

$$\Delta U_{\mathrm{int}} = K_{\mathrm{i}} - K_{\mathrm{f}}$$

$$m_{\mathrm{cb}} c_{\mathrm{cb}} \Delta T_{\mathrm{cal}} = \frac{m_{\mathrm{m}} {v_i}^2}{2} - \frac{m_{\mathrm{m}} {v_{\mathrm{f}}}^2}{2} \Rightarrow \tag{10.1}$$

$$\Delta T_{\mathrm{cal}} = \frac{m_{\mathrm{m}}\left(v_{\mathrm{i}}^{2} - v_{\mathrm{f}}^{2}\right)}{2m_{\mathrm{cb}}c_{\mathrm{cb}}}$$

The students can evaluate the expression by checking the units and if the limiting cases make sense. Inserting the data listed above, they calculate $\Delta T_{\mathrm{cal}} = 4.2^{\circ}\mathrm{C} \pm \delta\mathrm{T}$. Students in introductory courses can determine the uncertainty of the result δT by using so called weakest link rule, i.e. using the largest relative uncertainty of a single quantity. In our case the largest relative uncertainty (about 11%) comes from the term $(v_{\mathrm{i}}^{2} - v_{\mathrm{f}}^{2})$, giving $\delta\mathrm{T} = 0.50^{\circ}\mathrm{C}$. This value is only slightly smaller than the more accurate value $\delta\mathrm{T} = 0.54^{\circ}\mathrm{C}$, which is caluclated using the well-known uncertainty propagation equation (also called error propagation formula). Finally, the students need to compare the calculated value $\Delta T_{\mathrm{cal}} = 4.2^{\circ}\mathrm{C} \pm 0.5^{\circ}\mathrm{C}$ with the value obtained from the IR image: $\Delta T_{\mathrm{IR}} = (23.3^{\circ}\mathrm{C} \pm 2.2^{\circ}\mathrm{C}) - (21.6^{\circ}\mathrm{C} \pm 0.1^{\circ}\mathrm{C}) = 1.6^{\circ}\mathrm{C} \pm 2.3^{\circ}\mathrm{C}$ and discuss the results.

We find that $\Delta T_{\mathrm{cal}} > \Delta T_{\mathrm{IR}}$ and that the values do not match even if the uncertainties are taken into account. This is the moment to discuss how the assumptions that the students made affected the results and whether the invalidity of one or more assumptions justifies the discrepancy between the results.

If assumption 1 is not valid, then ΔT_{cal} should be larger than the measured value. If assumptions 2 and 3 are not valid, ΔT_{cal} should be smaller than the measured value. If assumption 4 is not valid, then ΔT_{cal} can be larger or smaller than the measured value, depending on which object has a higher initial temperature. Taking into account the precautions in performing the experiment (as explained above) and calculating the change in marble kinetic energy due to gravitational attraction students can conclude that the major source of discrepancies comes from assumptions 2 and 3 and that the sign of the observed discrepancy is consistent with the sign predicted based on their analysis.

10.2.1.3 How Do Students Respond to the Activities with an IR Camera?

How does observing the IR video of the experiment in addition to observing experiments in visible light affect students' ideas and reasoning about the experiment? Do infrared videos and images enrich students' experiences or do they not need to see IR data to reason about internal energy successfully? In order to answer these questions, we designed the following study.[1]

The participants were 46 18-year-old high-school students in Ljubljana, Slovenia, who already learned about work and energy as well as about the first law of thermodynamics. They were randomly divided in two groups. The first group (VIS, 21 students) only saw the video of the experiment described in Activity 1 (*Pulling*

[1] The results are part of a masters thesis in progress by Urška Nered at the Physics Education program, Faculty for Mathematics and Physics, University of Ljubljana, Slovenia.

a heavy object along a rough horizontal surface) obtained with a regular camera in visible light. The second group (VIS & IR, 25 students) saw the same video obtained with a regular camera and the video of the same experiment obtained with IR-camera. The students were asked the following two questions:

1. "Watch the video(s) and describe as carefully as you can what you observed." and
2. "Explain the experiments that you saw using energy and energy conversions. Indicate any assumptions that you made."

We analysed students' responses using the following coding scheme:

Code	Description of the criteria
0	The student does not mention pulling the object and/or increase of temperature and/or internal energy.
1	The student only describes that the person is pulling the object and that the temperature and/or some form of energy increases/appears. OR The student explains that the sum of the forces exerted on the object is zero and therefore no work is done on the object and no energy conversion occurs.
2	The student relates the friction force to the increase of the temperature and/or internal energy of the object and/or the surface, without explicitly mentioning work done by the hand on the object.
3	The student relates the work done by the hand (or energy provided by the hand) with the increase of the temperature and/or internal energy of the object and/or the surface.

The results of the coding (see Fig. 10.6) show that while the relative number of students (as the percent of the group) in category 3 in both groups are about the same, there are more students in categories 0 and 1 for the VIS group and more students in category 2 in the VIS & IR group. We ran statistical analysis using the Fisher exact probability test (two-tailed test) (Fisher, 1922). We found the p value to be 0.065 (here the p value shows the probability that these two distributions happened by random chance), which means that the groups are not significantly different statistically. However, we found differences examining closely responses of the students who formally belong to the same category.

Table 10.1 shows examples of student answers in each category. We see that the VIS & IR group brings up evidence such as the temperature change (or colour change in the video) as the reason for their explanations while the VIS group reasons mostly using their theoretical knowledge. (In the table, the parts of student explanations relating to the evidence are Italicized.) Therefore, we conducted the second analysis counting the number of students who mentioned experimental evidence while providing their explanations. We found that 7 students (30%) in the VIS group and 20 students (80%) in VIS & IR referred to evidence when providing explanations. This means that the students who had an opportunity to observe the IR video use evidence to back up their reasoning while the students who only saw the experiment in visible light reason using their prior theoretical knowledge.

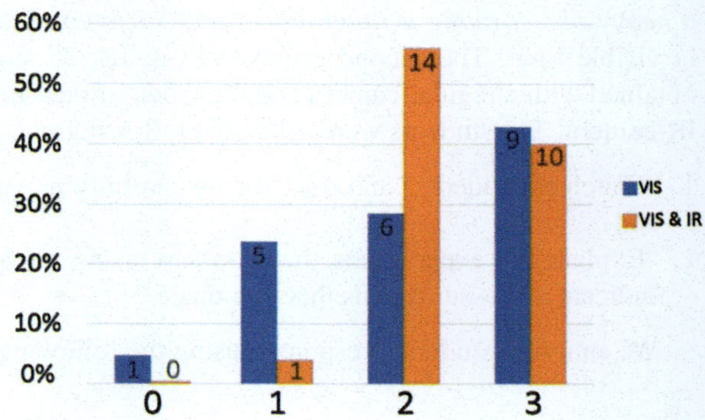

Fig. 10.6 Analysis of student responses. The numbers on the horizontal axis correspond to the code categories. The vertical axis shows the percent of students in each category, the numbers on the bars show the absolute numbers of students in each category

10.2.2 Learning How an IR-Camera Works

When students use an IR-camera as a black box, they do not need to know the details of its operation. In this section of the chapter, we describe two activities that help students construct a simple idea of IR imaging, which can then be expanded by the teacher (or by the students using additional resources) to explain the basic functioning of an IR camera (more precisely, a bolometer camera; see Chap. 2 of this book).

Through the activities described below, we want the students to learn the following ideas:

1. Every object, ("warm" or "cold") emits thermal radiation (EM waves) that can be focused using optical devices such as mirrors, to produce local changes in temperature on the screen that match the shape of the object – a thermal image of the object. In addition, every object ("warm" or "cold") also absorbs thermal radiation from the surroundings.
2. The changes in thermal radiation can be detected (converted into electric signals) using a small blackened thermometer.

From here on students can brainstorm how to apply these ideas to build a "real" IR-camera.

10.2.2.1 Activity 3 – Focusing Thermal Rays

Traditionally, students learn about mirrors using visible light. Working on the activity below allows students to expand focusing ability of a mirror to thermal radiation. To be successful, the students need to be familiar with ray diagrams, curved mirrors, and the basics of black body radiation.

Students watch the following experiment (see video https://youtu.be/u1iDY4WiGHg). An experimenter fixes a set of LEDs distributed in a shape of letter L next to a black screen (see Fig. 10.7) positioned in front of a concave mirror

Table 10.1 Examples of student responses from both groups coded in each category

Code	Example from VIS group	Example from VIS & IR group
0	It seems that the bob has a large mass; we can hear some rustling as the bob is being pulled.	No examples
1	We are pulling a cylinder across the table with a string. At the bottom of the cylinder is something strange. The hand does the work. The cylinder receives some kinetic energy. Then the work of friction force stops the bob.	In the first video we see only a bob that we move but we don't see any of the other changes that are visible in the second video. Energy of the bottom of the bob and of the part of table across which the bob moves changes. *We see this as a change of colour from green to yellow/orange.*
2	We pull a bob that has sandpaper at the bottom across smooth flat surface with a snare. Due to the bob having sandpaper at the bottom, the sandpaper warms when it chafes against the surface and its internal energy changes. The bob also has kinetic energy because it has velocity and it moves.	The bob is tied with a string, which we pull straight parallel to the surface. In the second video, we *can see that the bob leaves behind a trace where the temperature is higher than elsewhere on the table.* As we lift the bob up, we also *notice a different colour on the bottom* due to the bottom heating up while being pulled. Because the bob has some kinetic energy as it is being pulled and friction is slowing it down, part of this kinetic energy transforms into internal energy, which in turn heats the surface and the bottom of the bob.
3	I noticed that the energy had to be supplied so that the bob could move. *The bob left behind a trace where it sanded the wood.* To put the bob in the state where it was moving, energy – work – had to be supplied. As the bob was moving the friction force acted upon it – change in internal energy.	As the cylinder is being pulled both the surface and the cylinder itself warm up. *In the infrared video we notice that the bottom of the cylinder gets much warmer than the surface as the colour changes.* In the video we do work by pulling the cylinder. With this work the cylinder gains kinetic energy which than transforms into internal energy of the cylinder and surface due to friction, as a result the temperature increases.

The parts of student explanations relating to the evidence are Italicized

(data for our mirror: focal distance 16.5 cm, diameter 30 cm). The students need to describe what they see (a concave mirror, an object and a real inverted image of the object on the black screen) and to explain the outcome of the experiment using their knowledge of geometrical optics (Fig. 10.8).

The next experiment (see the video https://youtu.be/dMYR1iiwjEY and Fig. 10.9) uses the same setup as before, only this time, we fix a hot metal object painted black and shaped in letter L at the same place where the L-shaped LED object was before. Afterwards the students examine the screen and the object using an IR-camera (we used a FLIR One Pro camera attached to an Android smartphone).

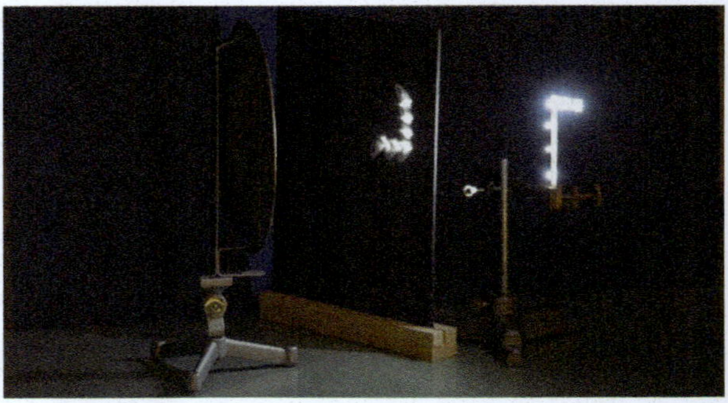

Fig. 10.7 An object (6 LEDs in shape of L) and its real image produced by a concave mirror. We expect the students to draw the ray diagram similar to Figure 10.8. They find that the image is indeed real and inverted

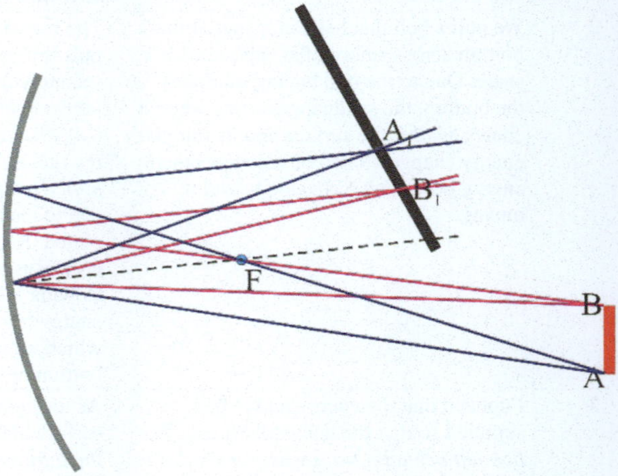

Fig. 10.8 An expected ray diagram

Students need to describe what they see and explain the outcome of the experiment. We expect the students to see an L-shaped region with increased temperature on the screen and to reason that the image on the screen is the real inverted image of the hot object.

Students might be able to suggest that the hot object emits thermal radiation (infrared electromagnetic waves) and the mirror focuses the radiation on the screen. As the result, parts of the screen become hotter. The infrared camera "sees" those parts – an inverted image of the hot object. It is important to encourage the students to think about energy processes that resulted in the screen becoming hotter. Until the experimenter fixed the hot object in the stand, the screen was in thermal equilibrium with the surroundings. Once the hot object had been placed in the stand, the temperature of the image parts on the screen started to increase because the net energy that those parts receive (due to the increased thermal radiation from the hot object) was larger than the net energy they gave away by thermal radiation, convection and conduction. As the temperature of the image parts on the screen increased, those parts started to emit more thermal radiation. Once the energy given away became equal to the energy received, the temperature of heated parts stopped

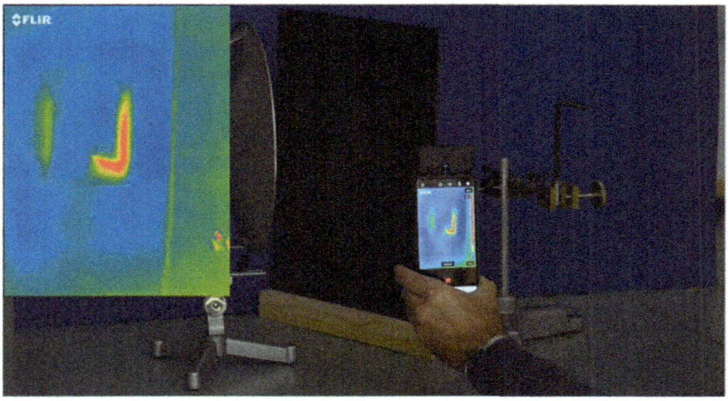

Fig. 10.9 A hot L-shaped object (black object on the right) and its thermal image produced by the concave mirror and examined by the IR-camera. The insert in the left top corner shows a magnified screen of the smartphone

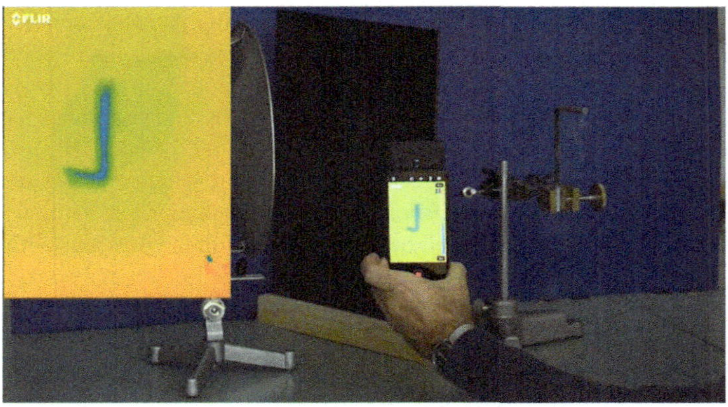

Fig. 10.10 A cold L shaped object (black object on the right) and its thermal image produced by the concave mirror and examined by the IR-camera. The insert in the left top corner shows a magnified screen of the smartphone

increasing, i.e. the image part of the screen reached equilibrium temperature that was higher than the temperature of the surroundings. Note, that in addition to radiation this process also involved thermal conduction and convection.

In the final experiment, the experimenter cools the black painted metal L (using liquid nitrogen, which temperature is −195.79 °C) and again fixes it at the same place as before (see the video https://youtu.be/LbonD73U1_E and Fig. 10.10). Students describe what they see and explain the outcome of the experiment. This time students will see an L-shaped region with decreased temperature on the screen.

Students are often surprised by this outcome but can reason to the explanation with a proper guidance from the teacher. Once the cold object is placed in the stand, the temperature of the image parts on the screen starts to decrease because the net energy that those part receive is smaller than the net energy they give away by thermal radiation, convection and conduction. As the temperature of the image parts on the screen decreases, those parts start to emit less thermal radiation. Once the energy given away becomes equal to the energy received, the temperature of cooled parts stops decreasing. The image part of the screen reached equilibrium temperature that is lower than the temperature of the surroundings.

10.2.2.2 Activity 4 – Testing Experiment: Using a Thermometer

After completing the previous activity, the students learn that mirrors can focus thermal radiation the same way as they focus visible light. However, as much as the comparison with visible light is helpful in the previous activity, it might also suggest that the thermal radiation *reflects* from the screen the same way this happens with visible light. In other words, the IR images that students observed when examining the screen do not necessarily activate the idea that the temperature of the screen at the image parts changes. There is another detail in the previous activity that can contribute to an unclear connection between the IR image and the temperature: the parts of the screen away from the image are blue ("cold") in the first experiment and yellow ("warm") in the second experiment. How can we explain this? Did the temperature of a screen (away from the image) change in each experiment? All issues mentioned above call for a new experiment in which we can observe the change of temperature at different points on the screen using a familiar device: a thermometer. The activity described below together with the previous activity allow students to devise a basic idea of how a bolometer-type IR-camera works.

If you ask the students to propose an experiment to test the hypothesis that the changes of the colour on the IR images of the screen in the previous activity are the result of the temperature changes of the screen, they might suggest to take two small thermometers and fix them on the screen, one at the place where the image appears and the other one elsewhere on the screen. The outcomes of these experiments are recorded in the videos https://youtu.be/xq1v0-n_-F0 and https://youtu.be/g6_nsnlKGFA and Fig. 10.11) and support the initial hypothesis. Additionally, the students notice that the changes in temperature, once we put the object in the stand, occurred with a significant time delay (about 1 min in our case).

In the experiments that we performed, we found that the image point temperature increased by about 2 °C when we used a hot (about 200 °C) object and decreased by about 1 °C when we used a cold (about −190 °C) object.

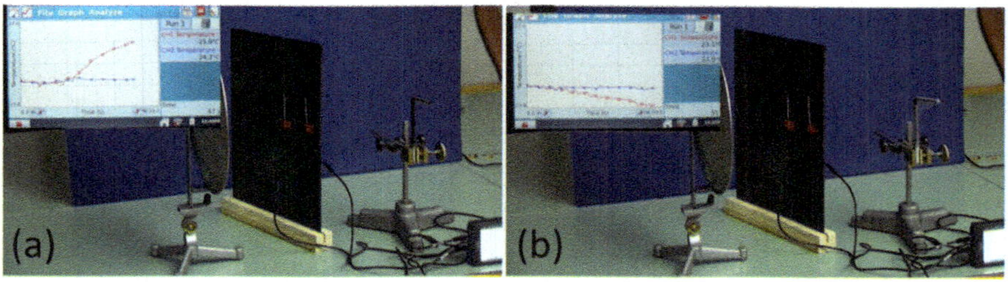

Fig. 10.11 Temperature-versus-time graphs as measured by two surface temperature probes (Vernier) when (**a**) using a hot object and (**b**) a cold object. The insert shows a magnified view of the data logger screen. In both experiments, the temperature scale is from 23 °C to 27 °C. The red curve shows the temperature reading in the screen area where the image of the object appears and the blue curve shows the temperature in the area away from the image

10.2.3 Learning New Physics Using the Knowledge of How an IR-Camera Works

The main idea of this component of the framework is to engage students in learning new physics topics by building on the knowledge about the IR-camera that the students developed through the activities described above.

The students will investigate how the emission of thermal radiation depends on the material properties of the object and through this investigation develop the concept of emissivity. The students will learn the following ideas:

1. Surfaces at the same temperature, but made of different materials or treated in different ways, can emit thermal radiation very differently. As the result, the temperatures of these surfaces as interpreted by an IR-camera can be significantly different compared to the actual temperatures of the surfaces.
2. Metal surfaces emit less radiation than most other surfaces at the same temperature while at the same time they reflect more radiation than other surfaces.

10.2.3.1 Activity 5 – Observing IR Images of Objects That Are Made of Different Materials

The activity can be framed as an investigation of how reliable the temperatures obtained by the IR-camera are. We have four different containers: a metal enamelled pot, a stainless-steel pot, a glass cylinder and a plastic 1-kg yogurt container (see Fig. 10.12). We pour about one litre of boiling water in all containers and then take IR images of each container (see Fig. 10.13). (All IR images in this activity were obtained with a FLIR ONE Pro camera and an Android smartphone and processed using FLIR Tools software.) Using IR images, we determine the surface temperatures of each container (small crosses in Fig. 10.13 indicate the spots at which measurements were done) and measure the temperature of the water in the

Fig. 10.12 Photos of different containers (from left to right): a metal enamelled pot, a stainless-steel pot, a glass cylinder, and a plastic container

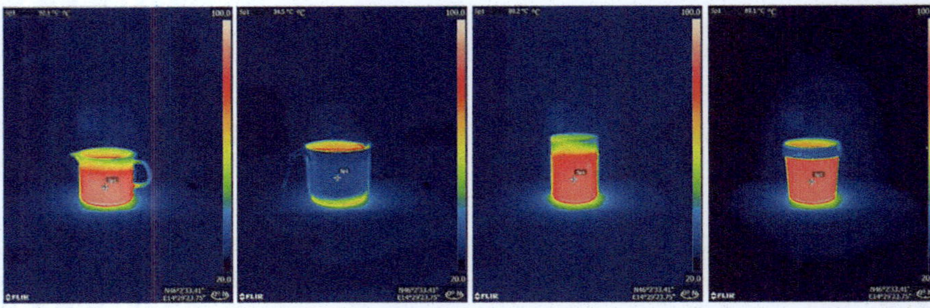

Fig. 10.13 The IR images of the containers in Fig. 10.12 after pouring boiling water into each. The colours on the images show the surface temperature of the objects in the interval from 20 °C (dark blue) to 100 °C (white)

containers and room temperature, using a conventional thermometer. All data are presented in Table 10.2.

We find that all IR images, except the image of the stainless-steel pot, show that the temperatures of the containers are much higher than room temperature and similar to each other. The students also notice that all temperatures of water obtained with the thermometer are close to the boiling point of water. The surprising data is the IR-based temperature of the stainless-steel pot. Based on everyday experience students know that the metal container will quickly reach the temperature of the water that we pour into the container. However, according to IR images, the temperature of the surface of the stainless-steel pot is more than 60°C lower than the temperature of the water in the pot. When students are asked to propose explanations, the following two ideas usually emerge.

1. *Explanation 1:* The temperature of the outside surface of the stainless-steel pot is (for some unknown reason) indeed equal to the temperature determined by the thermal camera.
2. *Explanation 2:* For objects made of certain materials (such as stainless steel) a thermal camera gives incorrect temperatures, while for other materials it gives correct temperatures.

Table 10.2 Relevant temperatures in the experiment with four containers filled with boiling water

	Metal enamelled pot	Stainless-steel pot	Glass cylinder	Plastic container
Temperature at the cross on the surface (determined from the IR image 30 s after pouring water)	92.1°C	34.5°C	89.2°C	89.1°C
Temperature of water (measured with a conventional thermometer, 30 s after pouring water)	93°C	93°C	94°C	96°C
Room temperature (measured with a conventional thermometer)	24°C	24°C	24°C	24°C

When asked to propose an experiment to test which explanation is better, the students usually suggest measuring the temperature of the surface of the stainless-steel pot with a conventional thermometer. We tried measuring the temperature using a Vernier surface probe but due to the poor thermal contact, we obtained the reading of about 60 °C, significantly lower than the temperature of water in the container. When we used a so-called heat sink compound (often used in electronics), we obtained readings close to the temperature of the water in the container. If we want to avoid discussion of thermal contacts at this point, we suggest the following qualitative testing experiment (sometimes it comes from the students): touch the surface of the stainless steel pot with a wax candle, which melting point is about 70°C. The outcome of the experiment (see Fig. 10.14a) shows that when we touch the stainless steel pot with a wax candle, the candle melts. Based on the outcome of the experiment, students can reject Explanation 1 but they cannot reject Explanation 2. If the students observe the same experiment, using an IR-camera, they will notice that melted wax looks significantly warmer than the metal surface of the pot (Fig. 10.14b). The students should realize that this observation is also consistent with the Explanation 2.

At this point, some students may propose to repeat the initial experiment, but this time we pour water in the containers that is cooler than the room temperature. Figure 10.15 shows IR images of the same containers after we poured ice-cold water into them. The relevant data are summarized in Table 10.3.

The students will again notice that all IR images, except the image of the stainless steel pot, show that the temperatures of the containers are much lower than the room temperature and that all temperatures obtained with thermometer are close to the

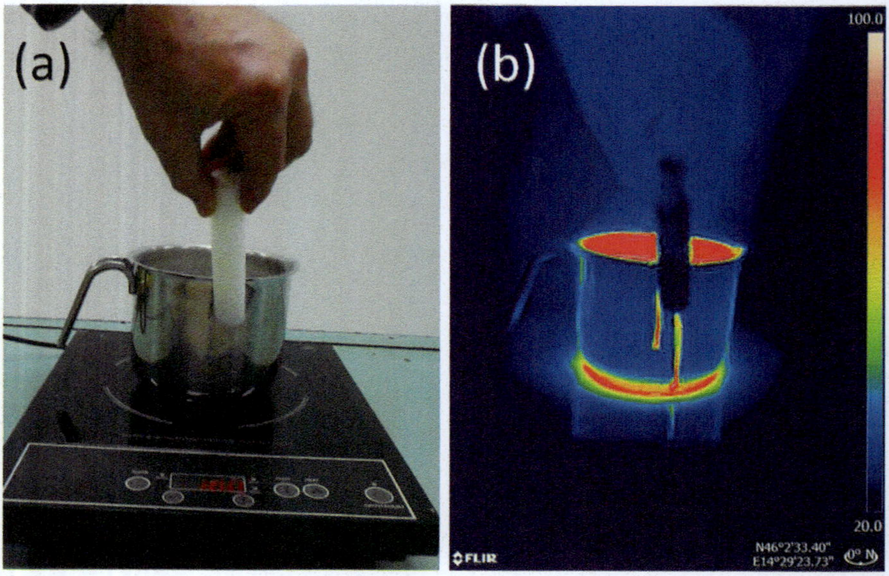

Fig. 10.14 Testing experiment with a candle: (**a**) visible light photo and (**b**) IR image. The colours on the IR image show the surface temperature of the objects in the interval from 20 °C (dark blue) to 100 °C (white)

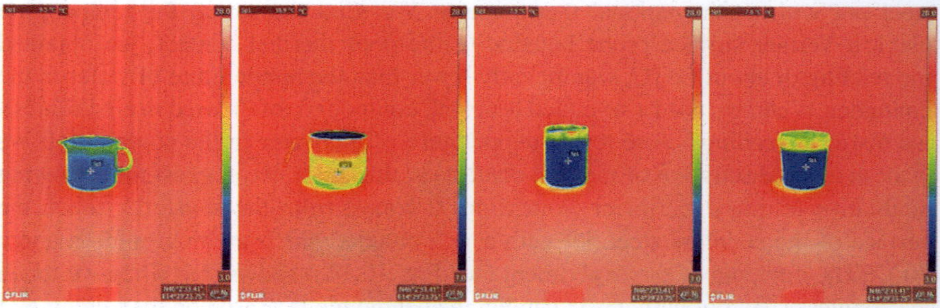

Fig. 10.15 IR images of the containers (see Fig. 10.12) after pouring ice-cold water into each container. The colours on the images show the surface temperature of the objects in the interval from 3 °C (dark blue) to 28 °C (white)

freezing point of water. When comparing the images and data obtained in this step with those in step a. students will first notice that this time the stainless-steel pot looks significantly warmer than other containers. Notice additional patterns:

1. *The temperature of the stainless-steel pot as interpreted by the IR-camera is close to the room temperature in both experiments.*
2. *The difference between the temperature of the stainless-steel pot as determined by the IR-camera and room temperature depends on the actual temperature of*

Table 10.3 Relevant temperatures in the experiment with four containers filled with ice-cold water

	Metal enamelled pot	Stainless-steel pot	Glass cylinder	Plastic container
Temperature at the cross on the surface (determined from the IR image 30 s after pouring water)	9.5 °C	22 °C	4 °C	5 °C
Temperature of water (measured with a conventional thermometer 30 s after pouring water)	3 °C	3 °C	3 °C	2 °C
Room temperature (measured with a conventional thermometer)	24 °C	24 °C	24 °C	24 °C

the pot surface. When the actual temperature is higher than room temperature, the surface appears slightly warmer than the surroundings and when the actual temperature is lower than room temperature, the surface appears slightly cooler than the surroundings.

Here the teacher helps the students devise an improved explanation:

An improved explanation: Metals emit less thermal radiation compared to other similar size objects at the same temperature. However, they reflect a larger part of radiation that comes from the surroundings than other objects. As a result, the temperature of the metal objects, as determined by the IR-camera, depends on the temperature of the object, on the material properties, morphology of the surface and on the temperature of the environment.

Therefore, once these parameters are known and set, the IR-camera gives the correct temperature of metal objects.

After completing the activities described above, the students are ready to learn about the new quantity *emissivity* ε. The teacher can introduce emissivity as a physical quantity that describes the effectiveness of a given surface in emitting energy as thermal radiation and present its formal definition. The emissivity is a ratio of the actual emitted irradiance of an object surface and the maximum possible emitted irradiance of the same surface at the same temperature. The latter is achieved when the object emits as a black body (see Chap. 2 in this book).

If this material is used in advanced courses, where students learn about Kirchhoff's law of radiation, the activity can be useful to prepare the stage for connecting/relating the idea of emissivity to the idea of reflectivity (Chap. 2). The experiments shown above offer even more opportunities to learn about new physics. Students might notice that the sides of the cylindrical containers on IR images (i.e. the tilted surfaces with respect to the observer) appear to have different temperatures than the part of the surface that is directly facing the observer. This observation can lead to the investigation on how the emissivity of different materials depend on the angle and to the introduction of the concept of Lambertian radiator (see Chap. 2 in this book).

10.3 Summary

The goal of this chapter was to show readers how one can use the advantages of an IR-camera in an introductory physics course without overloading the curriculum with new activities. To achieve this goal, we applied the framework for using modern devices in introductory physics courses (Planinsic & Etkina, 2019). So far, we have applied this framework systematically to light emitting diodes (Planinsic & Etkina, 2014, 2015a, b; Etkina & Planinsic, 2014) and to a light distance meter (Planinsic & Etkina, 2019). The IRT camera is our new contribution to the integration of modern devices into introductory physics.

The chapter describes five rich activities in which an IR-camera can be used as

(a) a black box to help students construct and quantitatively analyse the concept of internal thermal energy,
(b) a physics device whose operation students investigate and, through this process, deepen their understanding of electromagnetic radiation;
(c) a foundation to learn the concept of emissivity.

We propose experiments and describe in detail activities that students might conduct. We also include videos and photos for those who do not have an IR-camera or other equipment at their disposal. While the IRT cameras that attach to the phones are relatively cheap, they still might be out of reach for many teachers. Those who have an IR-camera will find our technical notes on how to conduct every experiment described in this chapter. Finally, we share our research on student responses to some of the activities.

We hope that this chapter will inspire the reader to not only try the activities described in this chapter but also take advantage of our work on modern devices in other areas of physics.

Acknowledgments The authors wish to thank Vida Kariž Merhar, Timotej Maroševič and Nada Žonta Kropivšek for helping us conduct research described in this chapter.

References

Daane, A. R., McKagan, S. B., Vokos, S., & Scherr, R. E. (2015). Energy conservation in dissipative processes: Teacher expectations and strategies associated with imperceptible thermal energy. *Physical Review Special Topics – Physics Education Research, 11*, 010109.

Etkina, E., & Planinšič, G. (2014). Light-emitting diodes: Exploration of underlying physics. *Physics Teacher, 52*, 212–218.

Etkina, E., Planinsic, G., & Van Heuvelen, A. (2019). *College physics: Explore and apply* (2nd ed.). Pearson.

Etkina, E., Brookes, D., Planinsic, G., & Van Heuvelen, A. (2020). *On-line active learning guide for college physics 2/e*. Pearson Education, Inc.

Fisher, R. A. (1922). On the interpretation of $\chi 2$ from contingency tables, and the calculation of P. *Journal of the Royal Statistical Society, 85*(1), 87–94.

Haglund, J., Jeppsson, F., Hedberg, D., & Schoenborn, K. J. (2015). Students' framing of laboratory exercises using infrared cameras. *Physical Review Special Topics – Physics Education Research, 11*, 020127.

Planinsic, G., & Etkina, E. (2014). Light-emitting diodes: A hidden treasure. *Physics Teacher, 52*, 94–99.

Planinsic, G., & Etkina, E. (2015a). Light emitting diodes: Exploration of new physics. *Physics Teacher, 53*, 212–218.

Planinsic, G., & Etkina, E. (2015b). Light-emitting diodes: Solving complex problems. *Physics Teacher, 53*, 291–297.

Planinsic, G., & Etkina, E. (2019). Framework for using modern devices in introductory physics courses. *European Journal of Physics, 40*, art. no. 065702.

Zull, J. E. (2002). *The art of changing the brain: Enriching teaching by exploring the biology of learning* (1st ed.). Stylus Publishing.

Chapter 11
Infrared Thermal Imaging: Applications for Physics, Chemistry and Biology Education

Choun Pei Wong and R. Subramaniam

11.1 Introduction

Especially in the past two decades, there has been so much of infusion of technology in the classroom. The classroom has moved away or is moving away from the chalk-and-talk model. Its operational efficiency has been significantly enhanced with the use of a computer, projector and network connection to the Internet. Use of PowerPoint presentations is common in classrooms in many developed countries, with access to the Internet helping to bring the rich resources of the web for use in teaching and learning. In more recent times, even mobile devices have been used to harness their affordances for teaching and learning (Lee et al., 2011). Generally, positive findings have been reported in the literature on the use of technology for teaching and learning of science. However, this technology generally excludes those that are of a specialized nature – for example, high speed imaging cameras and thermal imaging cameras. With their costs having come down in recent times, there is tremendous scope for the use of such tools in the science classroom.

Our focus is on the use of thermal imaging cameras for studying phenomena in Physics, Chemistry and Biology from a classroom perspective (Wong & Subramaniam, 2018a, 2018b, 2019, 2020a, 2020b). The motivation for our interest in this area stems from the observation that while there are a lot of research papers published in the medical and engineering fields on the use of thermal imaging cameras, it is less so in the field of classroom-based science. A possible reason could be the perception that such tools are expensive. Actually, costs have come down drastically over the years. A simple thermal imaging camera can now be purchased for about US$500, with a pre-owned version costing even less. Another reason could

C. P. Wong · R. Subramaniam (✉)
National Institute of Education, Nanyang Technological University, Singapore, Singapore
e-mail: chounpei.wong@nie.edu.sg; subramaniam.r@nie.edu.sg

J. Haglund et al. (eds.), *Thermal Cameras in Science Education*, Innovations in Science Education and Technology 26, https://doi.org/10.1007/978-3-030-85288-7_11

be the relatively low number of examples reported in the literature for their use in Physics, Chemistry and Biology classrooms. It could benefit from more examples. As it is, teachers are rather caught up with teaching, assessment and administrative tasks, all of which takes up a lot of time and that leaves them with little bandwidth to explore new tools for use in teaching. If a critical mass of examples on the use of thermal imaging cameras can be made available for use in the classroom on various topics in the sciences, this can be an incentivizing factor for teachers to consider using such tools to complement their teaching.

The infrared thermal imaging camera is one form of specialized technological device that has good potential for science education across all levels, from K-12 to university. Vollmer et al. (2001) have described its usefulness for science education as early as 2001. A number of other researchers have also developed applications for use in the sciences – for example, Xie (2011) in Chemistry; Short (2012) in the sciences; Melander et al. (2016) in Physics; and Haglund and Schönborn (2019) in Biology. A good discussion of the science of thermal imaging is given in Vollmer and Möllmann (2010).

Our motivation to enter the field of thermal imaging in science education was more to develop applications in Physics, Chemistry and Biology for use in the science classroom rather than to address any specific research questions. We hold the view that demonstrating the use of a thermal imaging camera for illustrating various phenomena in the sciences is in itself an endeavor worth pursuing. In this context, our work seeks to add on to the work of earlier researchers who laid the foundations for the use of thermal imaging cameras for exploring school-based science phenomena.

In this chapter, we illustrate the types of phenomena that can be explored using the thermal imaging camera in the sciences – mainly at the school level but also a few at the university level. This collection of ideas can be useful for the science classroom, laboratory or in the field, and can complement what is available in the literature. For each application developed, we also provide a few suggestions for further work.

All work reported here was done with a hand-held (1.05 kg) FLUKE Ti32 model. The infrared thermal imaging camera has a 320×240 focal-plane array, uncooled microbolometer detector. It can be used to explore phenomena over a temperature range of -20 °C to 600 °C, with an accuracy of ± 2 °C or $\pm$ 2%, whichever is higher. The sensitivity is ≤ 0.045 °C. The camera is powered by a rechargeable, lithium-ion battery. Infrared (IR) images captured by the camera can be analyzed using the SmartView software installed on a laptop computer. Setting of the emissivity values as well as the temperature range of the color bars can be done using this software. Emissivity is commonly defined as the ratio of the amount of radiation actually emitted from a surface to that emitted by a blackbody at the same temperature.

For convenience, we develop the chapter in terms of the three sciences.

11.2 Applications in Physics

In Physics, there are a number of phenomena or processes that can be better appreciated using the IR camera. We have explored concepts involving heat transfer and optical phenomena in this regard.

11.2.1 Cooling of Cans of Different Surface Finishes

To study the effect of surface finish of a can on the rate of cooling of hot water contained within it, a thermometer or temperature probe connected to a data logger is commonly used. Use of an IR camera is an alternative and provides a number of advantages – for example, the entire surface facing the camera can be studied for temperature changes during the cooling process and, when interfaced to a projection system, real time imaging is possible for students' view (Wong & Subramaniam, 2018a). Figure 11.1 shows the set-up for the experiment. Two aluminum cans of the same dimensions are used – one has a shiny surface while the other is painted dull black. Both cans contain hot water. As shiny metals have low emissivity values (between 0.05 and 0.07 for aluminum), and since materials with emissivity values of ≤ 0.2 are not easy to image using IR camera, a piece of small black electrical tape (emissivity value around 0.95) was stuck on the shiny can; it should attain the temperature of the shiny can rapidly. Another interesting observation of the IR image is that the reflection of the black can is more intense than the reflection of

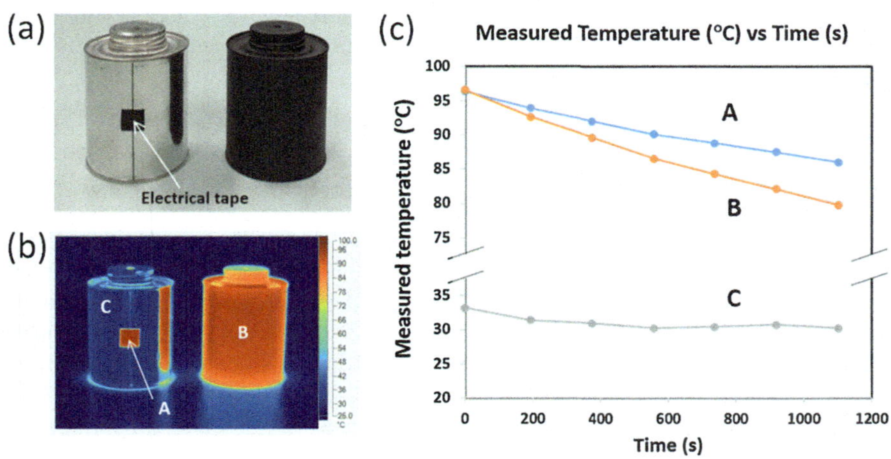

Fig. 11.1 (**a**) Light image of shiny can and darkened can. (**b**) IR image of same cans filled with hot water. (**c**) Temperature versus time graphs for shiny can (C), darkened can (B) and black electrical tape (A) on shiny can. © IOP Publishing. (Reproduced from Wong and Subramaniam (2018a) with permission. All rights reserved)

the shiny can – this is due to the black can being a much better emitter of radiation. Based on the IR images, the temperatures can be read off and a graph plotted. It can be seen that the temperature of the water in the darkened can decreases more rapidly than that of the shiny can because the former emits radiation more effectively and cools faster than the latter (Fig. 11.1).

Extension activities can include the following: using the same setup to explore the rate of warming of cold water inside the cans; using cans of different specific heat capacities and repeating the experiments with hot and cold water; using different liquids inside the cans and repeating the experiments at different temperatures; and varying the surface finishes of the cans and repeating the experiments. These would be suitable for middle school and high school students.

11.2.2 Work Done Against Friction

Work is done against friction when an object slides across a surface. This work is converted into heat. However, it is not easy to measure the temperature in such cases as the temperature change is rather small.

Our experiment involved rubbing an eraser across a tabletop (Wong & Subramaniam, 2020a). An IR image was taken immediately after the rubbing to ensure that the heat dissipation from the surface is minimized. A rise in temperature of 0.5 °C can be detected on the tabletop, as can be seen from the IR image shown in Fig. 11.2 – hardly discernible when resorted to via touch. A point to note is that it is difficult to demonstrate this effect using other temperature probes.

Extension activities can include measuring the temperature via IR images when the roughness or smoothness of the surface in contact with the eraser is varied and studying the same when the force applied to the eraser is varied. These would be suitable for middle school and high school students.

11.2.3 Infrared and Visible Light Transmission of Different Materials

The IR camera allows one to “see in a different light”, that is, it detects electromagnetic radiation in the infrared part of the spectrum, which is not visible to the human eye. We explored optical phenomena with two materials – silicon wafer and Perspex (Wong & Subramaniam, 2018a).

The silicon wafer (about 600 μm thickness) used for this experiment appears dark metallic to the eye. It cannot transmit visible light owing to its band structure. In the wavelength range of visible light, inter-band transitions occur via electrons (Fox, 2001; Palik, 1985) while in that for infrared light, lattice absorption bands are involved (Palik, 1985). The absorption and surface reflectivity are low enough, and

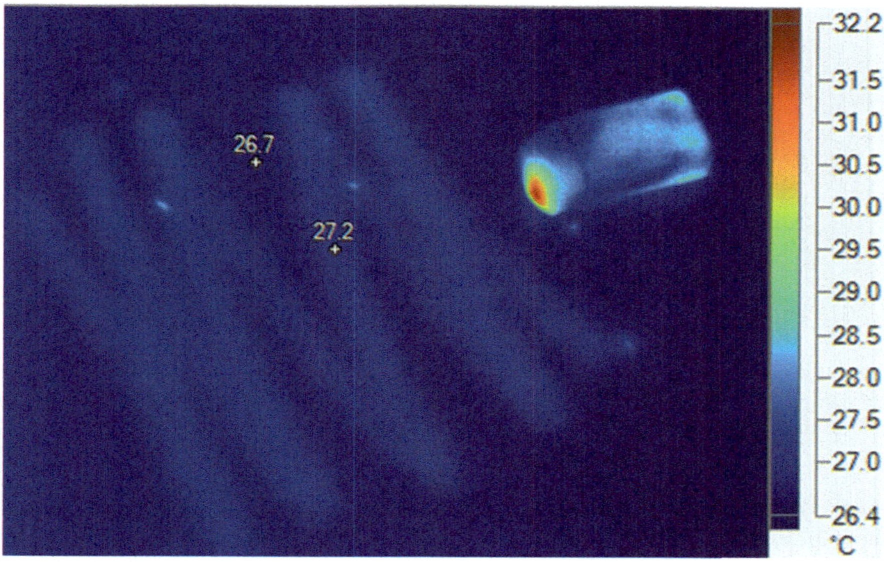

Fig. 11.2 IR image after rubbing eraser on tabletop. (Reproduced from Wong and Subramaniam (2020a) with permission)

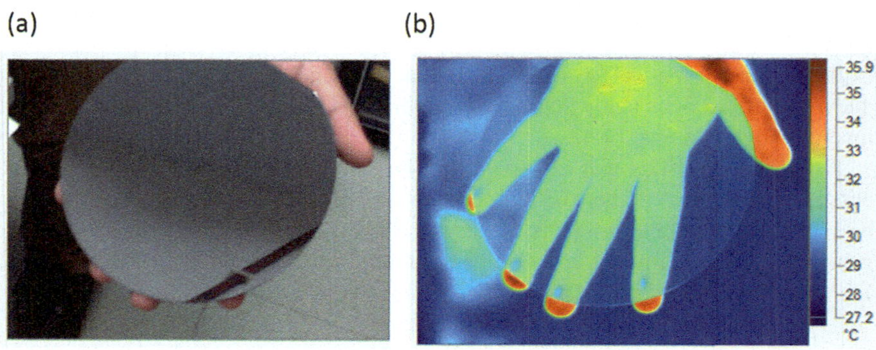

Fig. 11.3 (**a**) Light image of silicon wafer and person's hand. (**b**) Corresponding IR image. © IOP Publishing. (Reproduced from Wong and Subramaniam (2018a) with permission. All rights reserved)

the transmission high enough in the spectral range to detect heat from a person's hand through the wafer (Fig. 11.3b) via IR imaging. This can be contrasted with the image taken using a light camera, where the profile of the hand cannot be seen (Fig. 11.3a).

Extension activities can include using silicon wafers of different thicknesses and seeing whether the temperature distribution of the hand beneath it can be seen. It

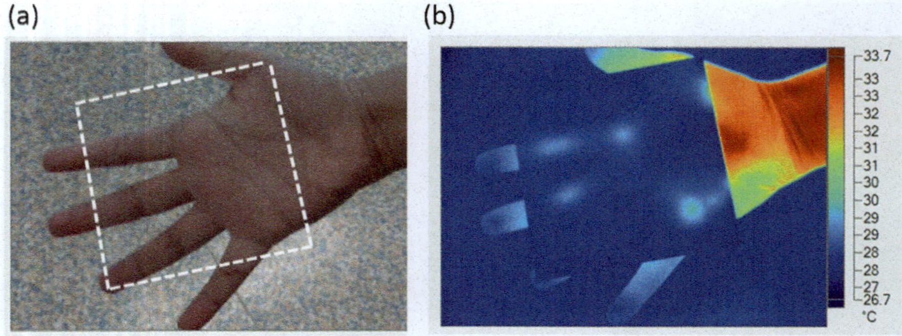

Fig. 11.4 (**a**) Light image of Perspex sheet and person's hand. Square outline is to show perimeter of transparent Perspex sheet. (**b**) Corresponding IR image. © IOP Publishing. (Reproduced from Wong and Subramaniam (2018a) with permission. All rights reserved)

has to be noted that the human hand is at a temperature of around 37 °C. Further activities can include using different objects of varying temperatures beneath the wafer and generating thermal images. These would be suitable for high school and university students.

The other material we used in our study was Perspex. Perspex does not absorb in the visible range of the electromagnetic spectrum, thus accounting for its high transparency in this region. At wavelengths corresponding to infrared radiation, absorption takes place, which results in low transmission. These optical behaviors explain what is seen in Fig. 11.4b. Note that a few hotspots seen in Fig. 11.4b are due to heat from the hand, which is warmer than the sheet. For comparison, the light image is shown in Fig. 11.4a.

Extension activities can include using Perspex of different grades and thicknesses, and seeing whether the temperature distribution of the hand beneath it can be seen. Using different objects of varying temperatures beneath the Perspex and generating thermal images can also be explored. These would be suitable for high school and university students.

11.2.4 Simple DC Electrical Circuits

When teaching simple electric circuits involving use of resistors (either as such or via incandescent filament bulbs), students are known to have difficulties (McDermott & Shaffer, 1992; Engelhardt & Beichner, 2004). Especially the ranking of bulbs in order of their brightness when arranged differently (series/parallel/etc) seems to be a problem. IR imaging can be used as a scaffolding tool to address this learning difficulty (Wong & Subramaniam, 2018b). This is because the brightness of the light bulb is proportional to its power. Along the way, students can also learn about joule heating. The more power that is dissipated, the higher is the temperature

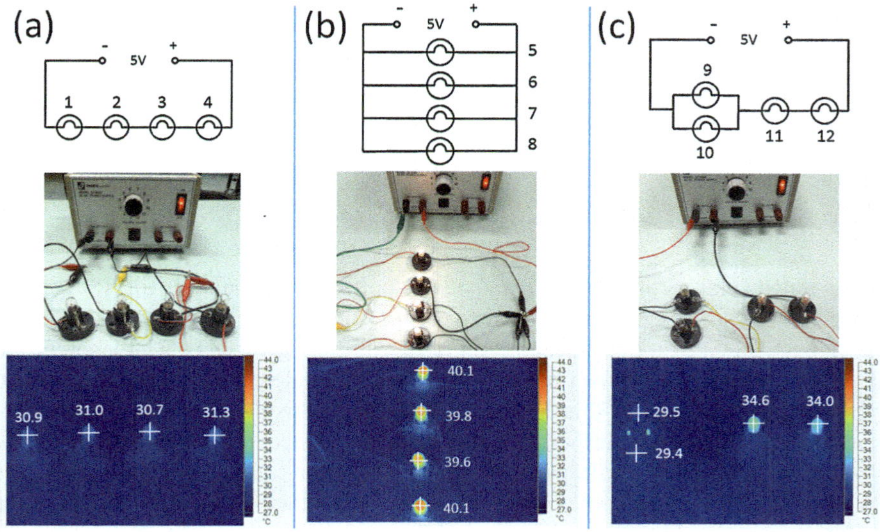

Fig. 11.5 Light bulbs in (**a**) series, (**b**) parallel, (**c**) mix of parallel and series combination. Top row: circuit diagrams, middle row: visible light images, bottom row: IR images. For reference, light bulbs are represented by numerals. Temperatures (°C) at corresponding locations indicated by + sign on IR images. © IOP Publishing. (Reproduced from Wong and Subramaniam (2018b) with permission. All rights reserved)

since there is greater heating. Netzell et al., (2017) have also reiterated that energy transformations in electric circuits can be better visualized using infrared cameras.

We used three different setups (Fig. 11.5). IR imaging directly shows the ranking of power dissipation of the light bulbs, all of the same wattage. There are also advantages over using a light meter to study this arrangement of bulbs since from a distance, it may not be that easy to discriminate among light intensities of bulbs that are only slightly different.

Extension activities can include the following: using bulbs of different wattage ratings in series and parallel combinations to generate IR images; using one bulb and progressively increasing the current so that the filament melts; using resistors instead of bulbs in the circuit; and working with further arrangements of bulbs to stimulate student thinking and then checking for verification using IR imaging. These would be suitable for middle school and high school students.

11.3 Applications in Chemistry

In Chemistry, there are quite a number of phenomena which can be better understood from a thermal imaging perspective. Criteria for inclusion include the need for temperature change in the processes. Quite a number of reactions in Chemistry entail temperature changes which are measurable.

Our work focused on developing applications, which have not been explored as well as those where a fresh perspective can be added.

11.3.1 Crystallization of Saturated Sodium Ethanoate

A saturated solution of sodium ethanoate in a bottle can be the basis for IR imaging. When a tiny seed crystal in solid form of this substance is added to the solution, it triggers crystallization – the solution turns white progressively. The temperature increases progressively with time and can be sensed if one holds the bottle. However, in a large lecture theatre, this can be an issue since it would be time-consuming to pass the bottle around. Further, if the bottle is passed around for students to touch, by the time it comes to the last student, there may not be sufficient heat to sense. This is where the IR camera comes with an advantage. As the process is exothermic, the solution becomes hotter and the heat is transmitted via conduction to the bottle. It is this heat change of the bottle that the IR camera senses. This temperature would essentially be the same as that of the contents in the bottle as the transfer of heat from the solution to the glass bottle should be rapid.

Extension activities using IR imaging can include the following: varying the size of the seed crystal to trigger crystallization and observing how the temperature distribution changes; and studying how cooling takes place once the contents in the bottle have reached its maximum temperature. These would be suitable for middle school and high school students (Fig. 11.6).

11.3.2 Heat Pack

Commercially available heat packs are used as therapeutic agents to relieve pain in parts of the human body. The active ingredient in the heat pack is a substance that exhibits metastable equilibrium between the solid and liquid states, and the transition from the latter the former can be effected by flexing a metallic disc that releases tiny particles which act as nucleation sites to promote crystallization. This is an exothermic process. That means, the temperature increases during the phase transformation. IR imaging presents an interesting way to show the temperature changes occurring during the process – starting from the nucleation site and spreading across the rest of the solution. Figure 11.7 illustrates this.

Extension activities can include the following: studying how cooling takes place once the contents in the heat pack have reached their maximum temperature; and studying the efficacy of different brands of heat packs. These would be suitable for middle school and high school students.

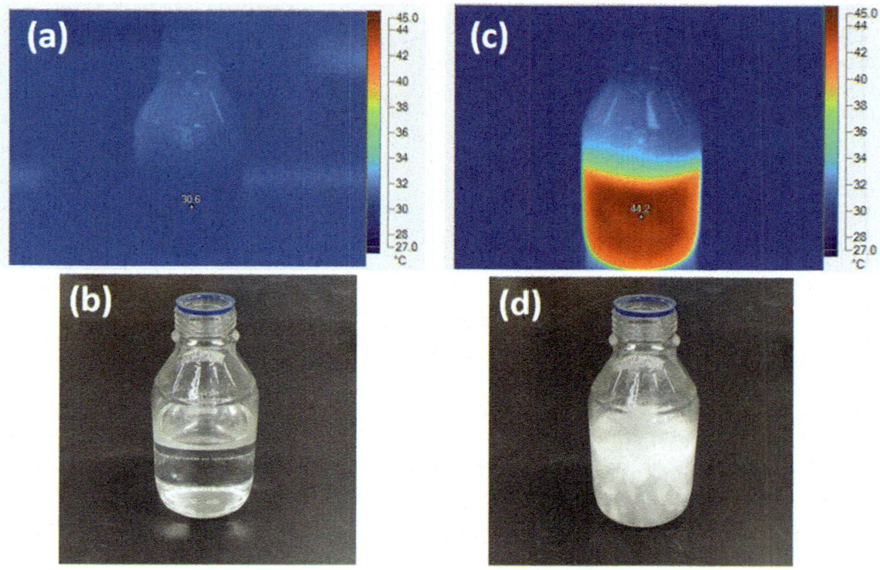

Fig. 11.6 Before crystallization: (a) IR image and (b) light image. After crystallization: (c) IR image and (d) light image. (Reproduced from Wong and Subramaniam (2020a) with permission)

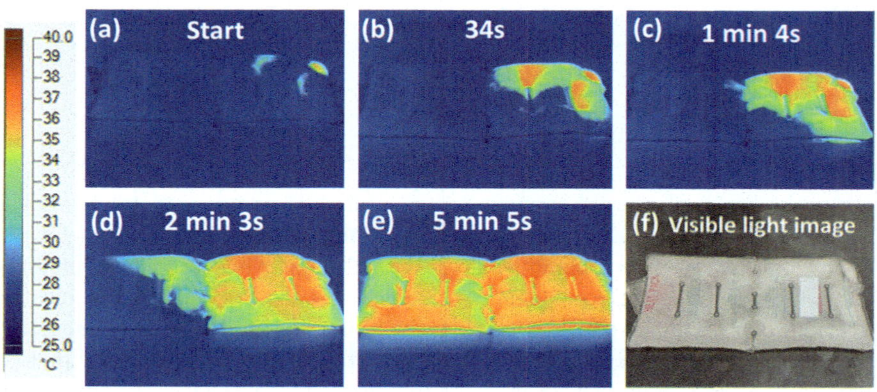

Fig. 11.7 (a–e): IR images during process of crystallization; color bar on left is temperature scale. (f) Light image after contents in pack have fully crystallized. (Reproduced with permission from Wong and Subramaniam (2019). Copyright (2019) American Chemical Society)

11.3.3 Endothermic Process

When glucose dissolves in water in a beaker, the temperature of the solution decreases, and the beaker is cold to the touch. This is an endothermic process. The more the amount of glucose dissolved in the solution, the greater is the

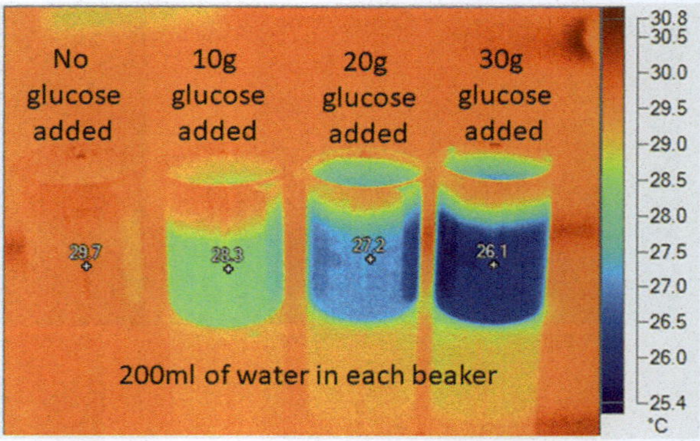

Fig. 11.8 IR images with different amounts of glucose dissolved in water in beakers. Color bar on right shows temperature scale. (Reproduced with permission from Wong and Subramaniam (2020a))

lowering of the temperature but only up to a certain limit. For this work (Wong and Subramaniam, 2020a), we used four beakers of water, each containing 200 cm^3 of water. Varying amounts of glucose were dissolved in it (0 g, 10 g, 20 g and 30 g). IR images taken after complete dissolution shows that as the molarity of the glucose solution increases, the temperature becomes progressively lower (Fig. 11.8).

Extension activities can include the following: studying how warming takes place as a function of time once the contents in the solution have reached its lowest temperature; and studying the endothermic processes of different substances in solution. These would be suitable for middle school and high school students.

11.3.4 Cold Pack

Ice packs are used as cold compresses for injuries incurred in sports as well as to reduce fever. We used a commercially available disposable cold pack for this demonstration (Wong & Subramaniam, 2019). The outer pouch contains urea and ammonium chloride while the inner pouch contains water. An endothermic reaction occurs when the inner pouch is ruptured by squeezing it. This causes the active ingredients to come into contact with the water. The process is facilitated by shaking the entire pouch vigorously to promote mixing of the reactants. The dissolution process is endothermic. IR imaging of the cold pack shows lowering of temperature of up to about 10 °C to 12 °C within 2 min (Fig. 11.9b).

Extension activities can include the following: studying how warming takes place once the contents in the cold pack have reached their lowest temperature; and studying the efficacy of different brands of cold packs. These would be suitable for middle school and high school students.

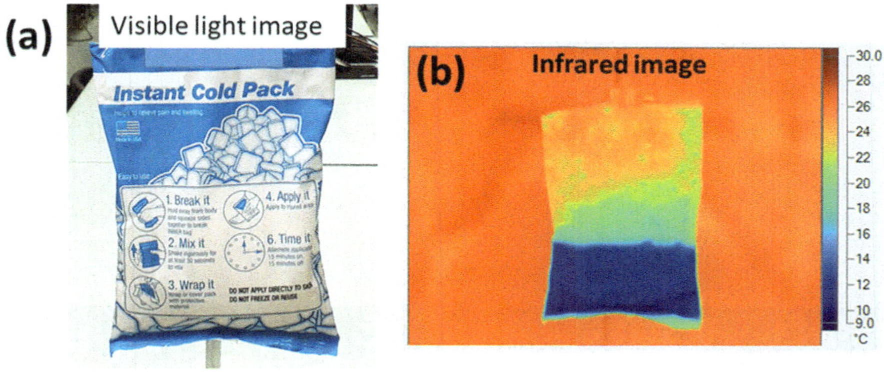

Fig. 11.9 (**a**) Light image of cold pack. (**b**) Corresponding IR image. (Reproduced with permission from Wong & Subramaniam (2019). Copyright (2019) American Chemical Society)

11.3.5 Heat Blocking Gel

The demonstration was inspired by the work of Kuntzleman et al. (2013) on fire gels. The fire gel they developed was able to provide a ring of protection for a hot dog when exposed to an alcohol flame. More specifically, they coated a thermocouple with this gel and set alcohol on flames in a pie tin. We feel that that IR imaging would be less messy in terms of subsequent clean-up (Wong & Subramaniam, 2019). For making the gel, we followed their procedure and obtained a transparent colorless gel. It was used to coat one side of a craft wooden stick (Fig. 11.10a), which was then held above a flame (Fig. 11.10b) for 30s. The coated side faced the flame. IR image of the top side of the craft stick (Fig. 11.10c) shows that the gel has been very effective in acting as a thermal insulating layer to reduce the transmission of heat from the fire to the other side of the craft stick. Figure 11.10 illustrates this.

Extension activities for IR imaging can include the following: studying the efficacy of different brands of heat blocking gels using the same experimental setup and exposing the gel-coated material to progressively higher temperatures and testing its efficacy as a heat insulator. These would be suitable for high school and university students.

11.4 Applications in Biology

The subject of Biology provides ample opportunities for using IR cameras to study live species. Humans and animals have a certain body temperature, and this can be the basis for exploring a myriad of contexts in Biology. In the case of plants, the IR camera comes in handy when studying thermogenesis. The portable nature of the hand-held IR camera also makes it ideal for use in fieldwork, especially for

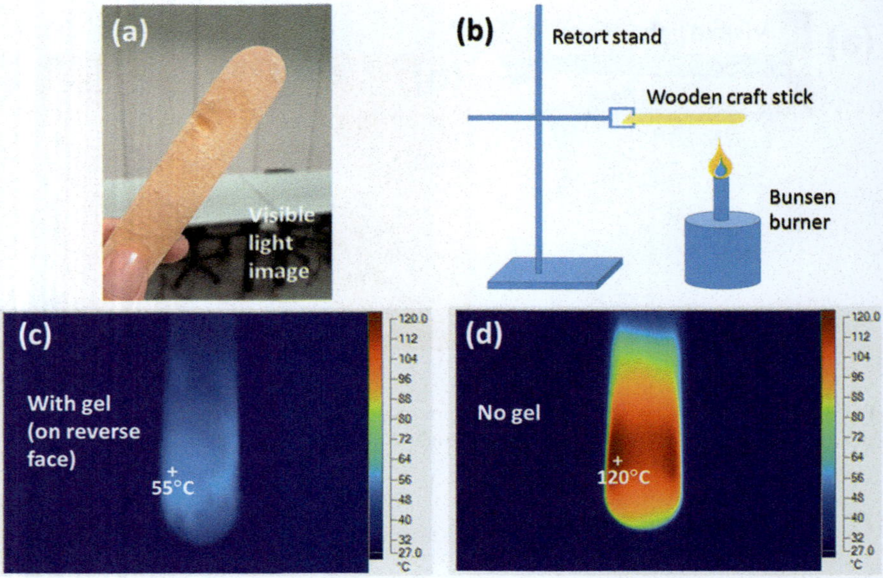

Fig. 11.10 (**a**). Wooden craft stick coated with transparent gel. (**b**) Experimental setup. (**c**) IR image of craft stick coated with gel. (**d**) IR image of craft stick without gel. (Reproduced with permission from Wong and Subramaniam (2019). Copyright (2019) American Chemical Society)

studying plants in ponds and animals in the wild. Moreover, the fact that imaging can be done from a distance allows even dangerous animals as well as plants nestling in the middle of a pond to be studied. This reduces stress levels of the species being studied. Here, we have to note that when traditional temperature probes are used, the animal subject may have to be restrained, and this poses issues in taking IR images under stress-free conditions. Compared to traditional temperature probes that can monitor only a particular location in the species effectively, IR imaging allows for the entire surface of the species facing the camera to be temperature-mapped.

11.4.1 Cold-Blooded Animals

In cold-blooded animals (poikilotherms), the body temperature varies with the ambient temperature, unlike warm-blooded animals (homeotherms) where the temperature is carefully regulated through homeostasis. A good example of cold-blooded animals is the common frog. It would thus be of interest to explore a demonstration of the cold-blooded nature of an animal under IR imaging (Wong & Subramaniam, 2020a). We chose the common frog in a frog farm for this purpose. The frogs in the farm are kept in their pen and are partially submerged in water (Fig. 11.11). Water allows very little infrared radiation to pass through, so the submerged

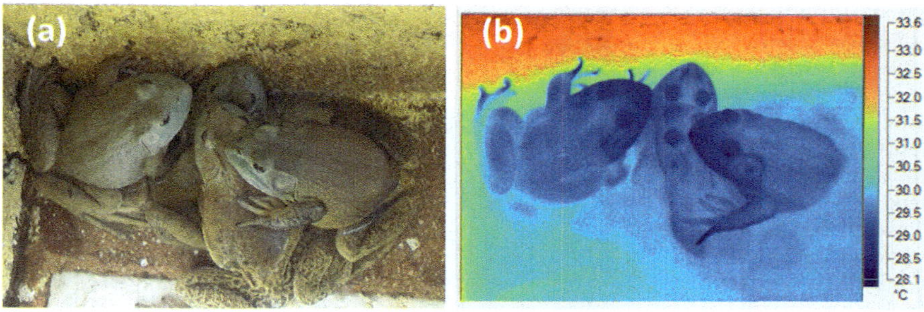

Fig. 11.11 Frogs in their enclosure (**a**) Light image (**b**) IR image. (Reproduced with permission from (Wong and Subramaniam (2020a)

part of the frog would not be accessible for IR imaging. On the day we shot the IR images, the side of the pen showed up as being warm since it was facing the sun. It can be seen that the temperatures of the entire bodies of the frogs are similar to that of the water they are partially immersed in (Fig. 11.11b). A light photograph is placed beside for comparison (Fig. 11.11a).

Extension work can involve studying other cold-blooded animals such as lizards, snakes and other species in their habitats. These would be suitable for middle school and high school students.

11.4.2 Temperature Distribution Across a Bird

IR imaging which was done in a bird park in Singapore illustrates how species can be studied from a distance and without disturbance. We chose the Australian King Parrot which was nestled in a large walk-in aviary (Fig. 11.12) in the local bird park (Wong & Subramaniam, 2020b). Temperature variations can be seen across the body of the bird. The warmest regions are located near the eyes and feet while the cooler regions are the feathered parts at the tips of the wings and tail, findings which are supported by the literature. It has to be noted that the IR camera tracks the temperatures corresponding to the outer layers of the feathers rather than the skin temperature.

Extension activities can include studying body temperature distribution of various animals in the zoo, bird park and crocodile farm. These would be suitable for middle school and high school students.

Fig. 11.12 IR image of an Australian King Parrot. Bar on right shows temperature scale. (Reproduced from Wong and Subramaniam (2020b), copyright © 2020 Royal Society of Biology, reprinted by permission of Taylor & Francis Ltd., http://www.tandfonline.com on behalf of Royal Society of Biology)

11.4.3 Thermogenesis

Here, we employ IR imaging to study thermogenesis in plants (Wong & Subramaniam, 2020b). Thermogenesis is the ability of plants to produce a significant amount of heat. The produced heat is used to evaporate chemicals contained within the plant so that these can act as scents to draw pollinators, create a thermal region for pollinators, or protect sensitive regions in the flowers (Seymour, 1997). IR imaging has previously been used to study thermogenesis in certain plants (Lamprecht et al. (2002).

It was not easy to find a plant which exhibits thermogenesis in the country. The literature indicates that a few species of the Nymphaeaceae family of water lilies are thermogenic. We managed to locate a species of this family in the botanic garden in the country. In the Symphony Lake in the Garden nestled various species of water lilies. IR imaging done remotely in the morning indicated that a particular water lily could be thermogenic (Fig. 11.13). Certain regions of the flowers show some hot spots in the IR image, and temperature differences of 3 °C could be noted. It was not a random occurrence as it was also seen at other times of the day. Other species in the lake did not show thermal gradients in the IR images. The particular water lily which seems to show thermogenesis is called *Nymphaea* 'King of Siam'. This finding has not been reported in the literature but confirmation using alternative approaches is necessary.

Extension work can involve locating and studying other species of plants which exhibit thermogenesis as well as surface distribution of temperatures in these plants. These would be suitable for high school and university students.

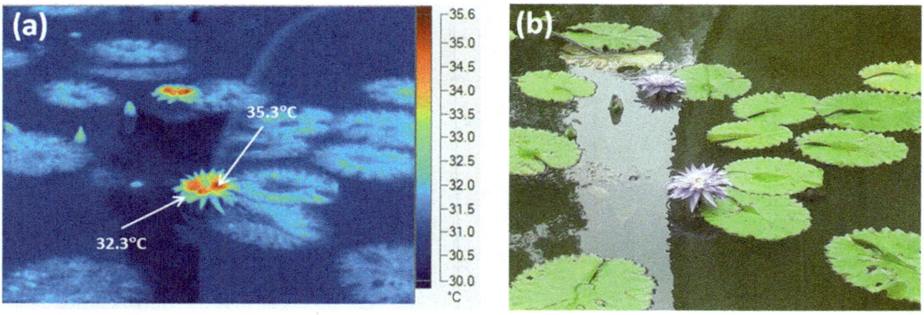

Fig. 11.13 (**a**) IR image of *Nymphaea* 'King of Siam' in a pond. (**b**) Corresponding light image. (Reproduced from Wong and Subramaniam (2020b), copyright © 2020 Royal Society of Biology, reprinted by permission of Taylor & Francis Ltd., http://www.tandfonline.com on behalf of Royal Society of Biology)

11.4.4 Humans

The use of IR cameras in airports to track visitors with fever is common in many countries. Since such imaging is very common, we focused on other aspects of human biology.

11.4.5 Hand Temperature Under Warm/Cold Conditions

Human beings are warm-blooded species. Textbooks often indicate that normal body temperature is about 37 °C. What has to be noted is that the skin often does not measure 37 °C. Temperatures in different parts of the body can show slight variations. IR imaging can be used to study skin temperatures in a non-invasive manner. Noteworthy is the point that an IR image of the human body facing the camera can show temperature profiles of the skin across the body, and this is evidence of temperature variations across the outer surface of the body.

In this demonstration (Wong & Subramaniam, 2020b), we used a setup where one hand is submerged in a tub of warm water maintained at 45 °C while the other hand is not. The submerged hand was removed after 5 min and dried quickly with a towel. An IR image was taken immediately. The temperature of the hand is about 40 °C, quite different from the normal body temperature (Fig. 11.14). Further, this demonstration shows that the process of heat conduction from the warm water to the hand is quite fast. That is, the hand gains heat in the process.

We repeated the same demonstration using cold water. One hand was immersed in a tub of cold water maintained at 17 °C while the other hand was not. The hand was removed from the cold water after 15 s, quickly dried with a towel and an IR image taken. The hand that was submerged in cold water was about 5–8 °C

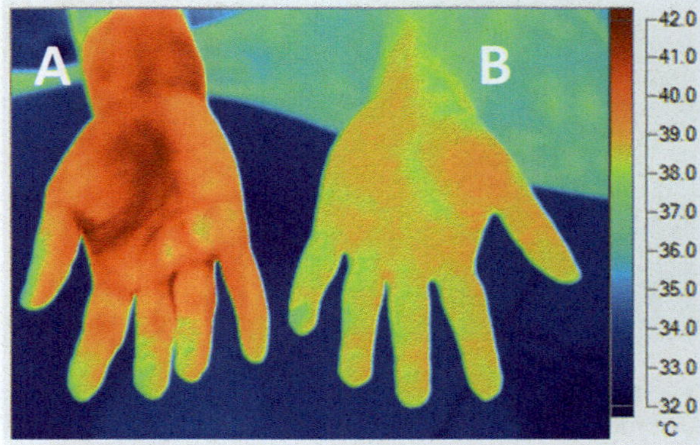

Fig. 11.14 IR thermal image of hand (**A**) in air after being immersed in water at 45 °C for 5 minutes and of hand (**B**) in ambient air. (Reproduced from Wong and Subramaniam (2020b), copyright © 2020 Royal Society of Biology, reprinted by permission of Taylor & Francis Ltd., http://www.tandfonline.com on behalf of Royal Society of Biology)

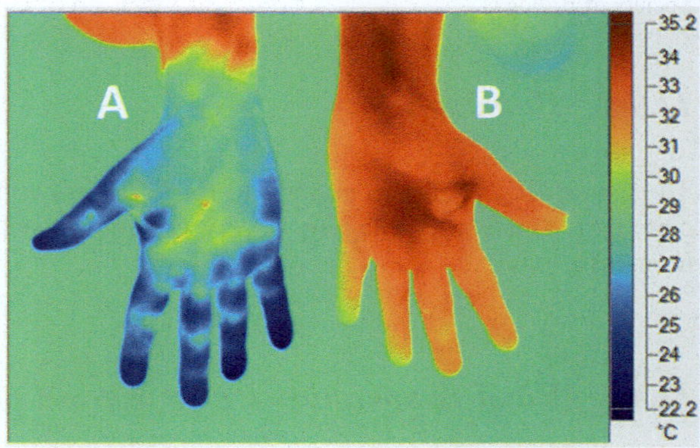

Fig. 11.15 IR thermal image of hand (**A**) in air after immersion in water at 17 °C for 15 s and of hand (**B**) in ambient air. (Reproduced from Wong and Subramaniam (2020b), copyright © 2020 Royal Society of Biology, reprinted by permission of Taylor & Francis Ltd., http://www.tandfonline.com on behalf of Royal Society of Biology)

cooler than that of the hand which was not submerged (Fig. 11.15). This simple demonstration also shows how fast the skin can lose heat in cold conditions.

Extension activities can include varying the immersion time of the hand in water of different temperatures and generating IR images. Care should be taken to ensure that the temperature is at a tolerable level.

11.5 Conclusion

The aim of this chapter was to showcase some applications in Physics, Chemistry and Biology where the use of the IR camera can confer useful insights from an educational perspective. These are also subjects in schools. As can be seen, quite a number of concepts in the sciences can be better appreciated using the IR camera than with the use of traditional temperature measuring devices or without these.

As mentioned earlier, our work in developing applications in the sciences was predicated by the need to build on the work of earlier researchers who have laid a foundation for the use of IR imaging in science education. The suite of applications that we have developed adds on to the repertoire of applications reported in the science education literature.

A point of interest is that we have suggested extension activities for each application that we have developed. This provides opportunities for further work as well as fostering the development of new applications of IR imaging.

Inquiry has been mainstreamed into science subjects in recent times. However, the kind of inquiry work that is generally done in schools is more of the type that needs to be done in the laboratory. While this will continue to be the mainstay of inquiry science, there are a number of investigations that can be done using the IR camera, both in the laboratory and in the field. These can provide novel experiences for students, and can be enriching.

In our work reported in the journal literature, we have consistently argued for the need for IR cameras to be brought into the science classroom or lecture theatre for teaching purposes. A number of concepts in the sciences can benefit by being presented from a thermal imaging perspective so that students can better understand the nuances of these concepts – for example, when teaching exothermic processes and endothermic processes, the scaffolding provided by the technological affordances of the IR camera can provide a different perspective for students to understand these processes. Just as web-based animations, simulations and YouTube videos are being used in the classroom for the teaching of the sciences, there is also a case for the thermal camera to be brought into the classroom for certain topics in the sciences.

An argument that can be advanced against the use of the IR camera in the classroom or the laboratory is the cost. Contrary to popular perception, the cost of IR cameras has come down drastically over the years. As mentioned earlier, a version for teaching purposes can be bought for about US$500, which is very affordable for an educational institution. A second-hand version is even cheaper. In recent times, even mobile phone-based apps have come on board but resolution can be an issue though it would still be adequate for quite a number of applications. Moreover, a school purchasing an IR camera can avail it to teachers in all the three sciences (Physics, Chemistry and Biology). During their spare time, teachers can also be encouraged to develop further applications. When it comes to project work, the use of this technology tool by students opens up new possibilities to look at science investigations. The lifespan of an IR camera is typically about 10 years. That means the upfront cost can be amortized over this period very affordably. When this is considered in tandem with the benefits that the IR camera offers for teachers and students in the sciences over the years, the return on investment can be very significant.

We also need to be mindful of some of the limitations of IR cameras. For example, underwater processes cannot be explored since water absorbs much IR radiation. Also, it is hard to explore materials with very low emissivity values using IR cameras.

References

Engelhardt, P. V., & Beichner, R. J. (2004). Students' understanding of direct current resistive electrical circuits. *American Journal of Physics, 72*(1), 98–115.

Fox, M. (2001). *Optical properties of solids*. Oxford University Press.

Haglund, J., & Schönborn, K. J. (2019). The pedagogical potential of infrared cameras in biology education. *The American Biology Teacher, 81*(7), 520–523.

Kuntzleman, T. S., et al. (2013). Creating and experimenting with fire gel, an inexpensive and readily prepared insulating material. *Journal of Chemical Education, 90*(7), 947–949.

Lamprecht, I., et al. (2002). Flower ovens: Thermal investigation on heat producing plants. *Thermochimica Acta, 391*(1–2), 107–118.

Lee, J. K., Lee, I. S., & Kwon, Y. J. (2011). Scan and learn! Use of quick response codes and smartphones in a biology field study. *The American Biology Teacher, 73*(8), 485–492.

McDermott, L. C., & Shaffer, P. S. (1992). Research as a guide for curriculum development: An example from introductory electricity. Part I: Investigation of student understanding. *American Journal of Physics, 60*(11), 994–1003.

Melander, E., Haglund, J., Weiszflog, M., & Andersson, S. (2016). More than meets the eye–Infrared cameras in open-ended university thermodynamics labs. *The Physics Teacher, 54*(9), 528–531.

Netzell, E., Jeppsson, F., Haglund, J., & Schönborn, K. J. (2017). Visualising energy transformations in electric circuits with infrared cameras. *School Science Review, 98*(364), 19–22.

Palik, E. D. (1985). *Handbook of optical constants of solids*. Academic Press.

Seymour, R. S. (1997). Plants that warm themselves. *Scientific American, 276*(3), 104–109.

Short, D. B. (2012). Thermal imaging in the science classroom. *School Science Review, 94*(346), 75–78.

Vollmer, M., & Möllmann, K.-P. (2010). *Infrared thermal imaging: Fundamentals, research and applications*. Wiley VCH.

Vollmer, M., et al. (2001). There is more to see than eyes can detect. *The Physics Teacher, 39*(6), 371–376.

Wong, C. P., & Subramaniam, R. (2018a). Seeing in a different light – Using an infrared camera to teach heat transfer and optical phenomena. *Physics Education, 58*(3), 035007.

Wong, C. P., & Subramaniam, R. (2018b). Use of thermal imaging for understanding simple electrical circuits. *Physics Education, 53*(6), 063002.

Wong, C. P., & Subramaniam, R. (2019). Exploring thermal effects and behaviors of chemical substances using an infrared camera. *Journal of Chemical Education, 96*(10), 2339–2344.

Wong, C. P., & Subramaniam, R. (2020a). The infrared camera as a pedagogical tool in science education. *School Science Review, 101*(376), 75–78.

Wong, C. P., & Subramaniam, R. (2020b). Use of technology in biology education – Case of infrared thermal imaging. *Journal of Biological Education*. https://doi.org/10.1080/00219266.2020.1808510

Xie, C. (2011). Visualizing chemistry with infrared imaging. *Journal of Chemical Education*, 88(7), 881–885.

Chapter 12
Visualizing and Exploring Heat in a Science Center

Karljohan Lundin Palmerius and Konrad J. Schönborn

12.1 Introduction

As represented by the collection of work in this book, the last fifteen years or so have seen multiple contributions demonstrating how infrared camera technology can be used to visualize challenging thermal concepts across various science education domains that include physics (Schönborn et al., 2014; Melander et al., 2016), chemistry (Xie, 2011) and biology (Haglund and Schönborn, 2019). Whilst most attention has been given to the role of thermal cameras at various educational levels in "formal" settings (e.g. classrooms or laboratories), limited writing has considered the application of thermal visualization in "informal" learning spaces such as science centers and museums. Our own observations reveal that although many public spaces showcase infrared imagery in exhibits—by positioning a thermal camera at a center's entrance, pointing a camera at different objects, or allowing visitors to view their thermal appearance on a large display—few have exploited the power of combining different visualization techniques with pedagogical activities to actively engage visitors with thermal science concepts.

In this chapter we describe the implementation and use of thermal visualization technology and related activities in a science center context. Specifically, the objectives of the chapter are to:

- present the design and implementation of an augmented reality system for real-time visualization and exploration of thermal processes.
- provide examples of hands-on activities for visualizing various thermal phenomena in a science center setting.

K. L. Palmerius (✉) · K. J. Schönborn
Linköping University, Linköping, Sweden
e-mail: karljohan.lundin.palmerius@liu.se; konrad.schonborn@liu.se

J. Haglund et al. (eds.), *Thermal Cameras in Science Education*, Innovations in Science Education and Technology 26, https://doi.org/10.1007/978-3-030-85288-7_12

- describe potential pedagogical opportunities for accessing and engaging thermal phenomena with the system.

12.2 Visualizing Thermal Phenomena in Science Centers and Museums

Our review of the literature has revealed various descriptions of the use of thermal cameras and thermal imagery in conjunction with science center and museum visits. For example, as part of a flipped learning module titled “Hot or Not?”, that connected classroom teaching to a science center visit, Andersen et al. (2020) have reported how lower secondary school students explored the regulation of human body temperature by visualizing surface temperature on a large screen in a Danish science center. The study found that students used the camera imagery to test hypotheses of temperature differences and in a set of observations themed the “curious science student”, participants used the thermal camera projection as a natural component of their explorations of skin temperature.

In an earlier study conducted ten years ago, Atkins et al. (2009) investigated the role of a thermal camera exhibit on visitors’ exploration of the insulation properties of clothing at a museum of science in the USA. While visitors sat on a bench, they could simultaneously view a projected gray scale thermal image of their bodies on a screen. They also had nearby access to different types of gloves and mittens. The exhibit incorporated textual labels that included posing the question as to which type of glove visitors thought “insulated heat better”. Observations and analysis of video recorded conversations of interactions at the exhibit showed that visitors often spontaneously explored thermal properties of various objects divorced from those constrained by the textual label content. The varying selection of different aspects of the exhibit provided a hybrid experience for visitors to engage in scientific conversations about thermal insulation.

In our own work ((Larsson et al., 2019), we have observed how fourth grade pupils interacted with hand-held IR cameras while performing collaborative “thermal workshops” in a digital science center in Sweden. Activities included small student groups freely exploring the thermal properties of surrounding objects and structures, as well as engaging in a thermos modelling task with the aim to insulate a volume of hot water as efficiently as possible. Analysis revealed that the children expressed various metaphoric relations by interpreting heat, in terms of colour, movement and change, in their communication of thermal phenomena while interacting with the cameras. In this regard, the thermal camera provided a communicative meaning-making resource allowing for the generation and evaluation of hypotheses during the inquiry-orientated tasks. Whilst there was a pronounced conceptualisation of “heat” as a noun, the children often linked their uttered metaphorical relations with their sensory experiences of heat. The work indicates that providing experiences of multiple metaphoric relationships around space, colour, movement and change may offer meaningful resources for accessing and learning about the abstract concept of heat.

In parallel with advancing approaches to visualize thermal phenomena at the same public venue, in 2016 we developed an augmented reality system that allowed visitors to explore projected thermal imagery while physically manipulating real objects (Palmerius and Schönborn, 2016). A core component of this work involved conceptualising what visitor activities and thermal experiments could be conducted and intuitively visualized within typical room temperature ranges of a science center. As we describe in the subsequent section, this process culminated in further technological development of the system in tandem with developing an array of educational demonstrations and activities for exploring heat transfer.

12.3 An Augmented Reality System for Visualizing and Exploring Heat-Related Concepts

Our described system for interactive exploration of thermal processes (Palmerius and Schönborn, 2016) is based on Spatial Augmented Reality (SAR) principles. SAR was first introduced by Raskar et al. (1998) for describing Augmented Reality (AR) display systems that are not worn by the user, but instead positioned or mounted in a room. The SAR approach of interest here is based on projectors, projecting computer graphics onto walls, objects or table surfaces. Specifically, instead of overlaying the augmentation by displaying it in front of the real world, as is the case with hand-held or head-mounted solutions (e.g. Akçayır and Akçayır, 2017), this approach overlays the augmentation by projecting it onto the real world. By applying *projection mapping*, the associated software developed for the system can compensate for both parallax offset and the unevenness of surfaces, thus fully controlling the distribution of augmentation over the surface and how this should optimally relate to the viewer. Doing so potentially turns the full room into a 3D display system, as exemplified by the work by Benko et al. (2014). The use of structured light projection to extract surface geometries was cited as a technique to compensate for the shape of the projection surface, which has become one of the most common principles for depth cameras today (Bimber and Raskar, 2005).

In applying the SAR approach, our system projects the captured thermal image of an object onto the object itself: such as a hand, the table or any other surface, see Fig. 12.2. Since the visual information is provided on the surface of the actual objects, a single user, a pair or a small group can simultaneously perform activities and experiments both individually and in collaboration, with an even bigger group viewing the interactions without any additional display systems. Another application of SAR in public exhibition environments is the *Lake Visualization 3D project* (Woods et al., 2016), which offers opportunties for users to conduct hands-on experiments with freshwater lake ecosystems. In their exhibit, visitors model rivers, hills, lakes and canyons in a literal sandbox, while a projector displays topographical curves and how the modelled landscape results in flow and collection of water.

Prior to the development of our system, we encountered few examples of projected thermal data in the literature. In one example, a team of artists (Gladyszewski et al., 2013) posted a clip showing their "thermal video projection system", initially a part of the choreographic project *Corps Noir*. In this system they used a co-located thermal camera, video projector and video camera to create a live video feed that displays thermal imagery projected onto the bodies of dancers. In continuation of that work by Gladyszewski (2016), a modified projection was released that provided thermal augmentation of an interactively modelled clay surface. In an example closer to the thermal visualization context of our work, Kawamoto (2014) has presented in a blog and an online video what he terms the *ThermalTable*. In his system, a FLIR C2 camera is used to record thermal data from a table, data that are then "projected back" using a projector. To date, we cannot find any documented communicative, evaluative or educational initiatives related to these efforts.

12.3.1 System and Setup for Augmenting Thermal Imagery

As a SAR application, our system requires three devices: a thermal camera that processes detected thermal radiation of the objects on the table, a projector for displaying the associated thermal image upon the objects, and a depth camera for projection mapping. Due to parallax between the camera and the projector, this mapping is necessary for determining how to offset the thermal image in the projection so as to align it correctly with objects at different distances from the camera and projector. The current version of the system is equipped with a fan-cooled FLIR A35 camera, powered via an Ethernet cable that connects it to a computer, see Fig. 12.1. The depth camera in the system is a Kinect One model 1520 and the projector is a typical consumer model (Epson EB-1940W) with triple colour LCD shutters that ensure strong colour reproduction at native resolution.

Successfully setting up the system requires the three devices to be calibrated and carefully co-registered to accurately align the projected image with the imagery of the thermal camera. One approach is to adopt camera calibration techniques for all three devices, using thermal targets in alignment with projected images, a process described in detail by Palmerius and Schönborn (2016) and Borrmann et al. (2016).

Fig. 12.1 The three comprising devices of the system for projecting the thermal signature back onto the objects themselves, from left to right: depth camera (Kinect One), projector (Epson EB-1940W), mounted to project through a mirror, and a thermal camera (FLIR A35)

12.3.2 Concepts and Features Related to Visualizing Thermal Processes with the System

Application of pseudocolours is an established technique for improving contrasts and dynamic range in imaging and visualization of non-colour data, such as X-ray (Sheppard et al., 1969) or, in our case, thermographic data. Various colour scales are provided by thermal cameras for this purpose, for example "gray", "iron" and "rainbow". However, in SAR applications intended for educational purposes, the choice of colour scale has two further important roles. Firstly, since the projected colour will be added to the existing natural colour of an object of interest, it must be clear to the viewer whether the resulting colour stems from the object or from the augmentation. Secondly, the mapping between the captured thermal information and the colour needs to be as *intuitive as possible*, to improve visitors' understanding of what the colours represent and thereby correctly associate underlying thermal processes with what is visually communicated. In an effort to display temperature contrasts at the ranges where they might provide the most impact, we chose to use a narrow, fixed, static range focusing on deviations induced by thermal activities, instead of trying to communicate absolute temperature values.

To do so, we assume that these activities take place in room temperature so that objects and surfaces that are not involved in thermal processes also remain at room temperature; only objects that are involved in thermodynamic processes, adding or removing thermal energy, will have a temperature that deviates from the normal thermal equilibrium of the surroundings. By choosing the colour black, which does not project colour at all, for representing room temperature we get an effect of positive and negative thermal flow *lighting up* the surface. Objects and materials at room temperature are seen as they appear in the natural light of the exhibition, allowing the visitor to build a mental reference image of the work space, from which the augmentation deviates.

In some experiments it is advantageous to be able to also discriminate between "modest" and "extreme" thermal processes, allowing a temperature range deviation as large as ±10 °C from room temperature. To maintain this range, while at the same time enabling good discrimination even at lower deviations, we chose two colours to represent temperature increases and two to represent temperature decreases, with continuous gradients between these and the neutral black for any intermediate temperatures. Deducing four suitable colours for the continuum, however, was not a trivial task. For representing moderate heating and cooling, the colours of red and blue have natural associations in many cultures, such as relating red to "hot fire" and blue to "cold water". In our system, these colours are presented at ±5 °C with respective gradients either side of the black representation of room temperature. The two other colours that we require to portray extreme deviations on the scale, are not as easy to select—for example, human interpretation of the colour white is influenced by multiple experiences that include cultural utterances such as "white hot" temperatures, children's rhymes, sensory experiences of snow, coal amber, or even teachings of black-body-radiation. Nevertheless, we chose to represent highest temperatures with white, and connected the perception to the higher brightness of a "hotter fire", and represented lowest temperatures with cyan, where northern European communities might associate the colour with that of naturally occurring ice formations.

12.3.3 Perceiving, Visualizing and Communicating Thermal Processes Through SAR

From a cognitive perspective, we argue that interacting with the visualization system supports direct symbolic off-loading to the environment (Wilson, 2002). By applying an intuitively mapped colour visualization that is spatially aligned with the underlying processes, the need for active symbolic interpretation is minimized, compared to that required with, for example, a dynamic scale with legend, a point measurement, or a numeric temperature display. In addition, the system provides immersion, which has been shown to improve spatial understanding and natural interaction and manipulation (e.g. Bowman and McMahan, 2007).

In a meta-study, Akçayır and Akçayır (2017) analysed studies on the use of AR in education, and classified the advantages and challenges that were identified. Several studies showed that AR has the potential to enhance motivation, understanding and achievement, as well as decrease cognitive load. Interestingly, the more common negative aspects identified in the studies, such as usability issues, tracking related problems and other technical constraints, were not found to be associated with SAR, but rather with hand-held or head mounted AR systems. With SAR, such as the system presented here, there is no user interface that controls the software and no display system to hold or wear; interaction is only with the real objects of interest. Since the system is firmly mounted and pre-calibrated for where it is located, tracking is also less of a problem. Therefore, essentially the only usability issue to consider is the visual artifacts associated with SAR, such as shadows, real-time projection mapping and occlusion.

The standard colour scales used in commercial hand-held thermal cameras are dynamic and adapt to the recorded temperature ranges displayed on the screen. Although some devices allow the user to *lock* the scale so that it stops recalibrating to a new temperature range, it is still not possible to set this range. Thus, it is possible to see what objects are "warmer" or "colder", but absolute temperature can only be determined by perceiving the numerical value at a certain labeled point, or by interpreting the displayed colour in reference to a colour gradient legend. Although salient thermal conduction is clearly visible with such scales, we opine that calibrating against room temperature and using a colour scale directly connected to "positive" and "negative" thermal processes, instead of a dynamic colour scale that varies depending on what the camera is directed towards, has the potential advantage of providing a more nuanced understanding of less overt processes, or even provides a basis for discussing relatively "powerful" and "weak" thermal differentials. With the fixed scale adopted in our system, a learner can use the colour information alone to perceive how much above or below room temperature the surface of their hand, or any object of interest is. In fact, this equilibrium-related scale is also related to the psychophysics of human thermal perception: our tactile sense of temperature is not absolute, but relative (Jones and Berri, 2002); what we perceive as *warm* or *cold* is actually *thermal flow* to or from our skin, as a result of contact with an object or air being warmer or colder than our skin, respectively.

12.3.4 Augmented Hands-on Experiments for Visualizing Thermal Concepts and Processes

As has been shown in multiple teaching contexts, thermal visualization with infrared technology can be a powerful tool for demonstrating processes of heat transfer and conversion. In this section we describe how such processes and principles can be explored using the described augmented thermal visualization system. For ideal results in a science center setting, all these experiments should be conducted at room temperature and the objects involved should initially be in thermal equilibrium with the surroundings.

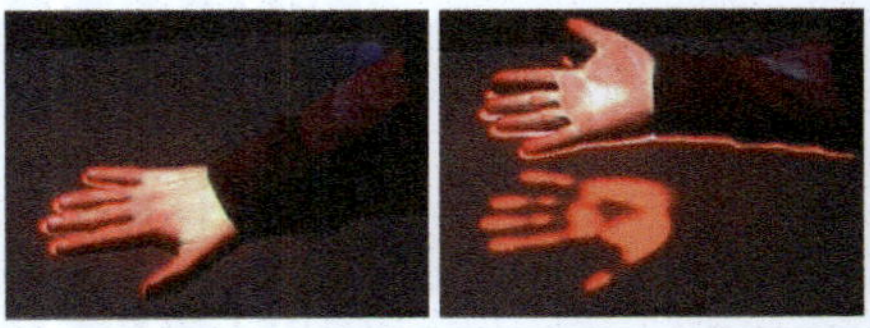

(a) Heat transfer from one's palm to the table surface upon thermal contact

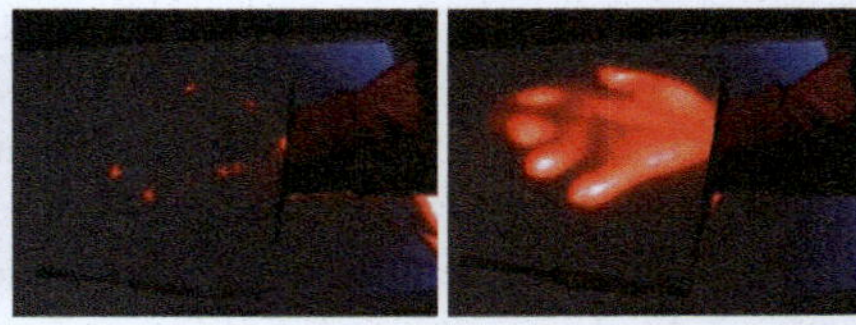

(b) Thermal conductivity is rapidly visualized when heat generated from one's hand penetrates a paper sheet

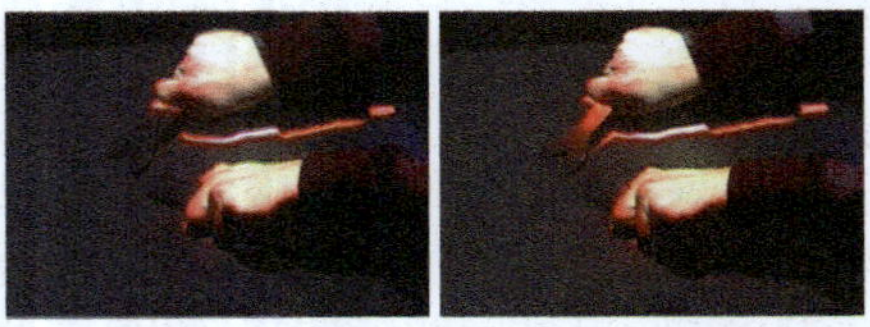

(c) Thermal conductivity of metals and insulators (e.g. wood) visualized during clasping the objects with one's hands

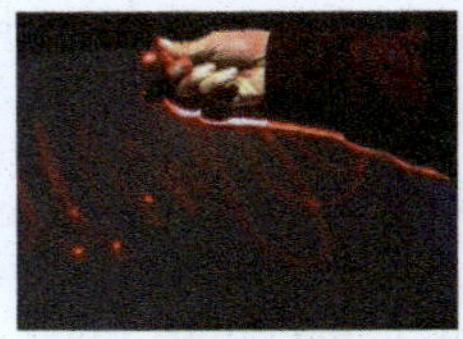

(d) Friction: Generating a heat trace upon sweeping an eraser across the table surface

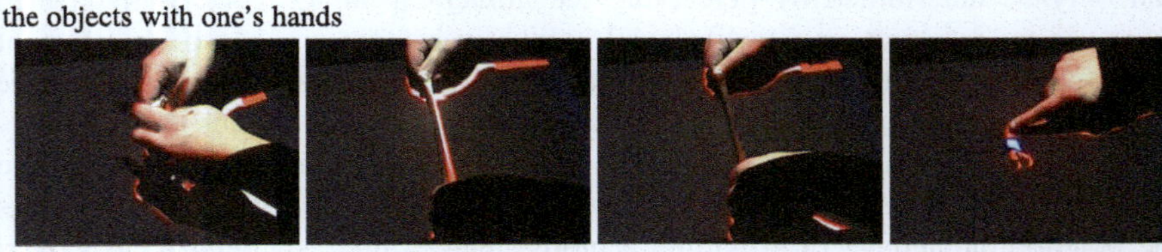

(e) Entropic forces: stretching a rubber band exhibits an increase in band temperature (red), followed by thermal equilibration to room temperature (black) upon maintaining band elongation, and concluded with contraction resulting in significant decrease in band temperature (blue)

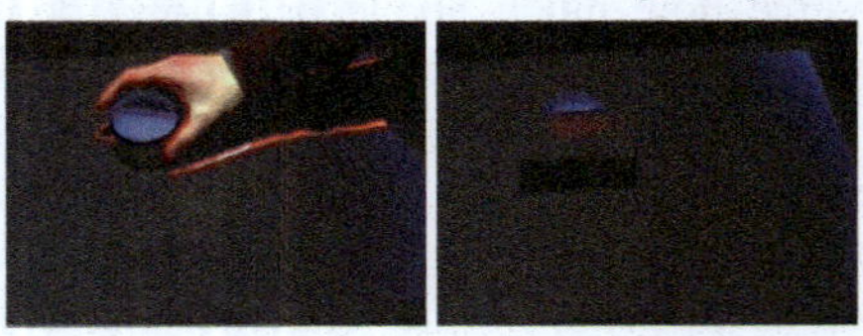

(f) Latent heat: the water surface is visualized blue due to evaporation, while the energy involved in the condensation process warms up the paper strip turning it red

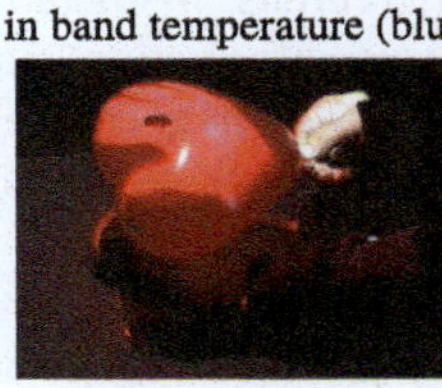

(g) Heat radiation wavelength: the balloon material is "transparent" to heat flow but not to visible light

Fig. 12.2 Photographs (**a–g**) capturing example demonstrations of using the augmented reality system in hands-on activities to explore thermal phenomena

Heat conduction can be demonstrated through numerous experiments, such as simply making contact with the table surface with one's palm. Upon lifting one's hand, the thermal signature of transferred heat is clearly visible by the highly contrasted palm imprint on the table. Upon subsequent flipping of one's hand, it is possible to see, albeit less clear than the table imprint, a slightly cooler palm surface as a result of thermal contact with the table surface, see Fig. 12.2a.

In another experiment, if one places one's palm beneath a sheet of paper (at room temperature), a thermal imprint of the hand rapidly appears, showing swift thermal conduction through thin insulators in a compelling explosion of colour, see Fig. 12.2b. This experiment can also be repeated using paper sheets of varying thickness/density for comparing different "rates" of thermal conduction. Also, by making contact with the surface of a stack of papers, such as a notebook, visitors can visualize the penetration depth of the thermal conduction by flipping through the pages and seeing a gradual diminishing of one's thermal palm print. By performing the same activity by placing both hot and cold objects on top of the paper pile, principles behind skin burns and the importance of extensive cooling of such burns can be demonstrated.

The comparison of heat conduction in elongated objects of different materials, such as rods is, while intuitive, not as powerful as one might initially think, see Fig. 12.2c. The heat transfer in thick metal is slow, as one has to heat up a considerable volume of metal, and a thin sheet will swiftly reach thermal equilibrium with its surroundings due to its large surface area, exhibiting dissipation rather than conduction. Therefore, a potentially more powerful indication of heat conductivity, albeit not very intuitive to a novice learner, is that the thermal imprint of a palm placed on top of a sheet will become less distinct and disappear faster the better the conductivity of the sheet. Producing an imprint at the boundary between two or more sheets of different materials thus provides a succinct visualization of the relative conductivity of different materials.

Heat Insulation, although essentially an aspect of heat conduction, can be more explicitly demonstrated in separate experiments that connect to its application and importance in everyday life. Indeed, when donning a glove and after waiting a short while, a visitor can use the system to distinctly see how the heat generated by their hand is "contained". In another experiment, although difficult for visitors to perform freely as part of a public exhibition, the slow mixing of hot and cold water can be a powerful demonstration of the insulative properties of water ($\sim$0.61 W/mK). The visualization shows distinctly red and blue regions, respectively, moving as one gently swirls the content of the container, thereby demonstrating that convection processes are more dominant than conduction in this experiment. Discussions of these demonstrations could be applied to the thermal conductivity of wetsuits and hydrothermal circulation of the oceans (e.g. Seager et al., 2002).

Friction can be demonstrated by using an eraser or another high friction object; as the visitor sweeps the eraser surface across the table surface a clear "heat trace" can be generated. While the heat generated on the table becomes distributed over the full length of the trace, it can also be clearly seen that the static hot spot on the eraser surface in prior contact with the table is at an even higher temperature, see Fig. 12.2d.

Impact heating can be demonstrated by striking a block of wood with a hammer. The wood undergoes plastic deformation while absorbing the kinetic energy of the strike, resulting in conversion to other forms of energy. The main energy component is transformed to thermal energy, which is visualized as a bright red mark on the dent.

Entropic forces are a thermodynamic property of both gases and polymers, such as those constituting a rubber band or balloon. The effect of entropy can be visualized when a rubber band increases in temperature as it is stretched (e.g. Pellicer et al., 2001; Melander et al., 2016), but is perhaps even more saliently demonstrated when it cools down upon contraction; if the rubber band is kept stretched until it reaches room temperature, upon subsequent contraction it will exhibit a temperature significantly lower than room temperature, see Fig. 12.2e, which in turn, serves as an indicator that rubber-band thermal dynamics cannot be explained by internal friction. The cooler areas are clearly visualized and can also be perceived by touching the contracted band or balloon surface with one's finger or lips.

This rubber band thermal effect can be compared with a gas state equivalent: provided with a sealed, metal bicycle pump, a visitor can compress air and see the container heat up, and upon releasing the pressure, see it cool down again. Consequently, both these experiments provide a basis for discussions on the thermal physics behind everyday technology such as a heat pump, fridge or freezer.

Latent heat, or enthalpy of evaporation and condensation, is demonstrated by partially covering a container of water with a thin piece of paper (e.g. Xie, 2011). The visualized evaporation event is an endothermic process, which is witnessed as cooling of the water surface (visualized as blue), while an exothermic condensation process occurs simultaneously on the underside of the paper, visually manifested as paper heating (visualized as red), see Fig. 12.2f. During the process the paper dampens, and even if not immediately discernible to the visitor, it can be demonstrated by swiftly flipping the paper on the table. When visualized with our system, the moist paper will transition from red to partially blue, as the moisture starts to evaporate, before transforming to black (room temperature) when all the contained water has evaporated.

Enthalpy of reaction can be visually communicated by dissolving different chemical compounds in water in respective endo- or exothermic chemical processes. Due to constant evaporation, the surface of water at room temperature equilibrium will be slightly colder than the surrounding air, which also serves as an additional reason for the extended dynamic range of the colour scale used in this work, as discussed in the section above. The water surface will always initially appear as light blue in the current science center environment. Introducing sprinkles of sodium carbonate into the water with mixing, results in an exothermic ("thermal energy releasing") process, which is visualized by the augmented red water. In contrast, dissolving ammonium bicarbonate results in an endothermic ("thermal energy absorbing") process visualized as a cooler cyan augmentation, which also represents our choice for pronounced negative temperature deviations.

Radiation wavelength properties can be represented by exploring the difference between visible light and thermal radiation, demonstrated with a thin insulator, such as an inflated balloon. When one's hand is positioned under an inflated balloon, the thermal signature projected through the balloon shows that the thermal radiation penetrates the balloon surface even though we cannot see through it with the naked eye, see Fig. 12.2g. By altering the volume of air contained in the balloon, visitors can study how the material thickness influences the transmission of thermal radiation.

12.3.5 Pedagogical Opportunities for Accessing and Engaging with Thermal Phenomena in a Science Center

The presented augmented visualization system offers various potential pedagogical opportunities for accessing and engaging with thermal phenomena in a public setting. We describe some observations of pedagogical opportunities and associated activities since implementation of the system in a science center setting.

Firstly, interacting with the system to access thermal phenomena caters to a *range of educational levels and age groups* from pre-school to university level. In this regard, it is possible to adapt many of the hands-on experiments described above depending on the target age or knowledge level. For example, we have observed preschoolers showing marked astonishment upon witnessing visualized heat conduction from one's hand positioned beneath a paper sheet. Such moments serve to provide an "entry point" for basic discussions about thermal concepts, and for generating further interest around deducing the "source" of the colour. At the same time, we have observed university students engage in discussions around statistical mechanics and thermodynamic properties (e.g. Roundy and Rogers, 2013) in relation to visualizing temperature changes associated with the stretching and contraction of a balloon or rubber band. Therefore, interacting with the system offers much *flexibility* in possibilities for performing specific (or combinations of different) experiments involving specific thermal concepts or principles. Performing these experiments can also be adjusted depending on the intended pedagogical aims of invoking interest in, engaging with, or developing conceptual understanding about thermal phenomena.

Secondly, we have observed multiple instances of *general public visitors'* engagement with the augmented visualization. Current printed text appearing beside the exhibit is purposefully brief, and reads, "Place your palm on the table, what do you see? What do you see when you lift your hand? Why?". We have observed that interacting with the system induces visitors' own self-initiated look-and-explore approaches (e.g. Schönborn et al., 2016a; Atkins et al., 2009), as well as setting the scene for additional scenarios where a center guide offers further opportunities to extend a visitor's engagement through dialogue. We have observed examples of such continued dialogue in the form of guides asking questions such as, "What do you think 'makes' the red colour", "Try rubbing that eraser across the table surface and see what happens", as well as explaining the hardware setup and functioning of the system, which many center visitors enquire about. Center visitors' experiences are also sometimes complemented with demonstrations of the hands-on activities by guides, as a means to invoke interest in, and further engagement with the system.

Thirdly, integration of activities with the augmented visualization has also featured in structured *school class visits* to the digital science center. As part of a broader aim of the center to promote problem-solving activities during out-of-school learning experiences, multiple workshops are designed to intertwine with the formal school curriculum. One example includes a set of thermal workshops developed by in-house science center pedagogues in consultation with school science teachers. One such workshop directed at fourth grade pupils involves various practical activities with hand-held thermal cameras in combination with student investigations, written questionnaires and discussions that include items such as, "What colours do you see when pointing the camera at different objects? What do these colours mean?", "Briefly touch the provided pieces of metal and wood. Direct the camera to the objects, how do they appear in the camera display?", and "Point the camera at one another, how does the appearance of one's face compare to one's clothes?" These activities are also followed by, or commence with, complementary demonstrations at the augmented visualization exhibit. In another workshop that was also the subject of a research study (Larsson et al., 2019), the sessions commenced with an introduction to IR camera technology and imagery using a large screen projection, while a researching pedagogue discussed the dynamic visualization. This was followed by various free exploration activities and a thermos modelling task, and culminated in a discussion with the facilitator. As a concluding pedagogical activity, students also had an opportunity to interact with the augmented visualization in their further hands-on interrogation of various thermodynamic phenomena encountered during the session.

Our observations of these activities highlight the potential role of the augmented setup when used alone, or in *combination with different technologies* (e.g. handheld IR cameras, large screen projections) to visually communicate thermal processes, activities that can also be implemented in *varying sequences* of access and engagement. Furthermore, the discussed pedagogical activities show a *diversity in potential pedagogical structure* for communicating thermal phenomena to visitors and students that include free exploration and discovery, guided demonstrations, and more focused workshop sessions.

12.3.6 Practical Considerations of the System for Public Exhibition Spaces

In comparison with the availability of thermal cameras in public science centers, the projected augmentation system described in this chapter exhibits a significant degree of robustness. The user of the system is not exposed to any electronics or other sensitive hardware equipment that could otherwise be easily damaged—these economically valuable features are mounted firmly and hidden from view. For the more advanced experiments that require more than merely placing one's hand on the augmented surface, there is a need for visitors to access additional materials. In the described activities we mention paper sheets, pieces of metal and wood, rubber bands, balloons, chemicals and liquids. Although fairly simple materials to procure, ensuring their availability and maintenance in a public setting requires personnel to regularly ascertain and monitor the materials, and see to the cleanliness of the exhibit space.

The application of projected SAR is associated with the disadvantage of the augmented visualization being affected by material surfaces. Dark materials darken the visualization display, and transparent and reflective materials distort the projection onto non-intended surfaces. Therefore, activity material for a projected augmentation system should exhibit a bright colour with low saturation, and preferably painted matte white. This also solves the problem of different measurements between objects with equal temperatures because of varying emissivity of material surfaces, a problem sometimes overlooked in thermal imaging. With paint covering the natural color and texture of objects, signs or labels will have to inform the visitor about the material. An alternative, however, is to make it an activity to guess the "hidden" material, for instance metal, glass or wood, from its weight and feel, and from its thermodynamic properties as visualized by the system.

To demonstrate thermal processes with our system, we emphasise visualizing how communicated changes in temperature deviate from room temperature. Therefore, any fluctuations in room temperature must be kept to a minimum. In the best scenario, external sensors could be used to allow the system to automatically monitor and compensate for unfavorable temperature fluctuations. Unfortunately, constant handling of surfaces and objects means that they heat up after only a few thermal demonstrations, making them glow red in the visualization environment and impeding the augmented visual impact of observing further heat processes. In addition, while many objects cool back down in a minute or so, it is a bigger challenge for bulkier material, such as the table top surface itself, to return to room temperature quickly enough, especially when there are frequent visitors interacting with the system. For this reason, positioning the exhibit in a well-ventilated space is necessary. Cooling of the table top surface can be further improved by integrating a thin metal heat sink in its construction.

Some of the described hands-on experiments include water and other liquids. Liquids are perhaps not the most favourable state of matter to freely explore in a public exhibit, and especially not in the context of an exhibit where even liquid droplets can disturb the overall thermal equilibrium of the table working surface. As a consequence, the liquid-based activities are more conducive to guided presentations or supervised demonstrations. Also, to better allow the projected visualization to appear on the liquid surface, we prefer milk and similar white liquids.

We also note that the experiments described in the previous section have been purposely customised for the current indoor projected augmentation system. Nevertheless, their conceptualisation and testing have also inspired thinking about other potentially powerful experiments that are beyond the immediate scope of the system described here, but that also constitute further out-of-school opportunities to visualize thermal phenomena. For example, directing a thermal camera at the

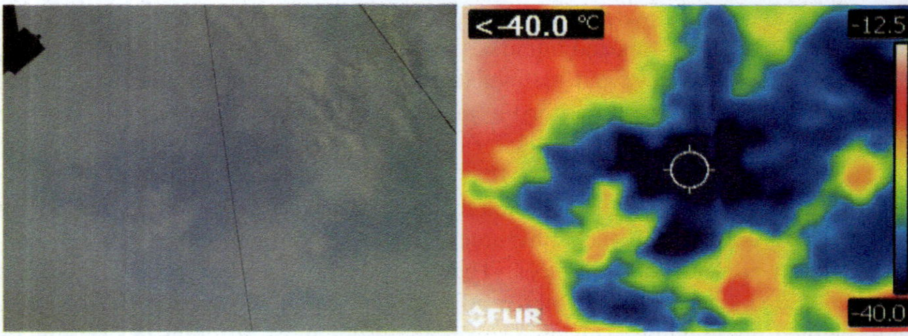

Fig. 12.3 Coloured photograph (left) and infrared thermograph (right) of a patchy, Scandinavian winter sky, taken with a hand-held thermal camera, providing a qualitative visual demonstration of radiative flux enhancement

sky provides a reading of the radiative flux towards the earth's surface, as radiated from what is in front of the camera—from clouds and air, and even from space—see Fig. 12.3. We duly acknowledge that although a typical hand-held thermal camera does not provide a mode for quantitatively measuring radiative flux, the qualitative visual comparison of the thermal reading of a cloudless patch of sky with that of cloudy areas can foster a discussion on the impact of air composition on the climate; in dry conditions, the measure given for the cloudless sky will be out of range (negative) for such cameras, and thus, since directing it at a cloud base provides an in-range reading (except for some clouds at high altitudes, see e.g. Szejwach, 1982) the clear difference in radiation from these patches provides a qualitative visual demonstration of the underlying principle of radiative flux enhancement (Walsh and Chapman, 1998). We also acknowledge that it is necessary to point out the limitations of the displayed thermal image during such qualitative demonstrations (see Chap. 9).

12.4 Implications of the Visualization System for Engaging Thermal Phenomena in Science Centers

This chapter has described how an augmented visualization system can be used in a science center context to expose visitors to thermal phenomena through multiple hands-on activities. In summary, visitors' interaction with the presented system in a public space allows for:

- Engendering interest and engagement about thermal phenomena through the power of interactive visualization (Schönborn et al., 2016a; Ynnerman et al., 2018).
- Combining interactive visualization with body-based and hands-on experiences to access abstract science concepts (Schönborn et al., 2016b; Höst et al., 2020).

- Generating and testing hypotheses about thermal concepts and principles and application to real world relevance (Larsson et al., 2019; Andersen et al., 2020).
- Offering potential connections between science center activities and school activities for learning about heat (Andersen et al., 2020; Larsson et al., 2019).
- Informing how interactive visualization exhibits representing abstract science content can be designed to satisfy intended communicative and pedagogical aims (Atkins et al., 2009; Hauan and Kolstø, 2014; Schönborn et al., 2016a).

From an educational research perspective, our future work in this area aims to further define what specific learning outcomes are intended with the augmented system activities, and then to empirically investigate what interactive processes could be related to visitors' acquisition of specific thermal-related knowledge. As pointed out by Hauan and Kolstø (2014), approaches that consider the nature and quality of pupils' physical explorative processes with exhibits in public spaces might help inform how "informal" activities can be more systematically connected with "formal" classroom learning of heat concepts.

A further perspective of high interest to our future research is the emerging paradigm of *exploranation*—a term coined by Ynnerman et al. (2018) that combines explanatory visualization (visualization for communicating scientific content) and exploratory visualization (visualization for exploring scientific data) in the active discovery of scientific knowledge. Our recent work (Höst et al., 2020) has applied the exploranation perspective in consideration of an interactive visualization designed for communicating nanoscience and nanotechnology to the public. The process resulted in testing and further specification of the theoretical tenets of exploranation, and their implications for visual communication and learning of science. The augmented thermal exhibit presented in this chapter combines interactivity, science communication and data visualization, the core components of exploranation. In conclusion, projecting dynamic thermographic data while simultaneously interacting with physical objects provides exciting and engaging pedagogical possibilities for learning about fundamental science concepts in public spaces.

Acknowledgments We thank FLIR Systems AB for the use of a FLIR AX5 camera and Anna-Karin Lindblom for supporting this work, *Visualiseringscenter C* for the opportunity to integrate the system as part of the public exhibition, and Andreas Larsson for discussions on using the system as part of pedagogical science center activities.

References

Akçayır, M., & Akçayır, G. (2017). Advantages and challenges associated with augmented reality for education: A systematic review of the literature. *Educational Research Review, 20*, 1–11. https://doi.org/10.1016/j.edurev.2016.11.002.

Andersen, M. F., Levinsen, H., Møller, H. H., & Thomsen, A. V. (2020). Building bridges between school and a science center using a flipped learning framework. *Journal of Museum Education, 45*(2), 200–209. https://doi.org/10.1080/10598650.2020.1744238.

Atkins, L. J., Velez, L., Goudy, D., & Dunbar, K. N. (2009). The unintended effects of interactive objects and labels in the science museum. *Science Education, 93*(1), 161–184. https://doi.org/10.1002/sce.20291.

Benko, H., Wilson, A. D., & Zannier, F. (2014). Dyadic projected spatial augmented reality. In: *Proceedings of the 27th Annual ACM Symposium on User Interface Software and Technology* (pp. 645–655). Association for Computing Machinery. https://doi.org/10.1145/2642918.2647402.

Bimber, O., & Raskar, R. (2005). *Spatial augmented reality – merging real and virtual worlds*. A K Peters.

Borrmann, D., Leutert, F., Schilling, K., & Nüchter, A. (2016). Spatial projection of thermal data for visual inspection. In: *2016 14th International Conference on Control, Automation, Robotics and Vision (ICARCV)* (pp. 1–6). IEEE. https://doi.org/10.1109/ICARCV.2016.7838617.

Bowman, D. A., & McMahan, R. P. (2007). Virtual reality: How much immersion is enough? *Computer, 40*(7), 36–43. https://doi.org/10.1109/MC.2007.257.

Gladyszewski, S. (2016). Argile et lumière—clay and light. https://vimeo.com/240518063. Accessed 4 Feb 2021.

Gladyszewski, S., Burton, A., Ricard, J., & Grenier, É. (2013). Live thermal video projection system. https://vimeo.com/60292952. Accessed 12 Feb 2016.

Haglund, J., & Schönborn, K. J. (2019). The pedagogical potential of infrared cameras in biology education. *The American Biology Teacher, 81*(7), 520–523. https://doi.org/10.1525/abt.2019.81.7.520.

Hauan, N. P., & Kolstø, S. D. (2014). Exhibitions as learning environments: A review of empirical research on students' science learning at natural history museums, science museums and science centres. *Nordic Studies in Science Education, 10*(1), 90–104.

Höst, G., Palmerius, K., & Schönborn, K. (2020). Nano for the public: An exploranation perspective. *IEEE Computer Graphics and Applications, 40*(2), 32–42. https://doi.org/10.1109/MCG.2020.2973120.

Jones, L. A., & Berris, M. (2002). The psychophysics of temperature perception and thermal-interface design. In *Proceedings 10th Symposium on Haptic Interfaces for Virtual Environment and Teleoperator Systems. HAPTICS 2002* (pp 137–142). IEEE. https://doi.org/10.1109/HAPTIC.2002.998951.

Kawamoto, K. (2014). Thermaltable. http://kenkawamoto-works.tumblr.com/post/106298696083/thermaltable-2014-dec-link-using-flir-one-an. Accessed 12 Feb 2016.

Larsson, A., Stafstedt, M., & Schönborn, K. J. (2019). Heat angels and paper cups: Pupils' use of metaphoric relations when engaging thermal cameras to investigate heat. In E. McLoughlin, O. E. Finlayson, S. Erduran, & Childs, P. E. (Eds.), *Bridging research and practice in science education: Selected papers from the ESERA 2017 Conference* (pp. 73–89). Springer. https://doi.org/10.1007/978-3-030-17219-0_5.

Melander, E., Haglund, J., Weiszflog, M., & Andersson, S. (2016). More than meets the eye–infrared cameras in open-ended university thermodynamics labs. *The Physics Teacher, 54*(9), 528–531. https://doi.org/10.1119/1.4967889.

Palmerius, K. L., & Schönborn, K. (2016). Visualization of heat transfer using projector-based spatial augmented reality. In L. T. D. Paolis & A. Mongelli (Eds.), *Augmented reality, virtual reality, and computer graphics*, Lecture notes in computer science (pp. 407–417). Springer. https://doi.org/10.1007/978-3-319-40621-3_29.

Pellicer, J., Manzanares, J. A., Zúñiga, J., Utrillas, P., & Fernández, J. (2001). Thermodynamics of rubber elasticity. *Journal of Chemical Education, 78*(2), 263. https://doi.org/10.1021/ed078p263.

Raskar, R., Welch, G., & Fuchs, H. (1998). Spatially augmented reality. In *Proceedings of The IEEE International Workshop on Augmented Reality*. A K Peters.

Roundy, D., & Rogers, M. (2013). Exploring the thermodynamics of a rubber band. *American Journal of Physics, 81*(1), 20–23. https://doi.org/10.1119/1.4757908.

Schönborn, K. J., Haglund, J., & Xie, C. (2014). Pupils' early explorations of thermoimaging to interpret heat and temperature. *Journal of Baltic Science Education, 13*(1), 118–132.

Schönborn, K. J., Höst, G. E., & Lundin Palmerius, K. E. (2016a) Interactive visualization for learning and teaching nanoscience and nanotechnology. In K. Winkelmann & B. Bhushan (Eds.), *Global perspectives of nanoscience and engineering education* (pp. 195–222). Springer. https://doi.org/10.1007/978-3-319-31833-2_7.

Schönborn, K. J., Höst, G. E., & Palmerius, K. E. L. (2016b). Nano education with interactive visualization. *Nano Today, 11*(5), 543–546. https://doi.org/10.1016/j.nantod.2015.10.006.

Seager, R., Battisti, D. S., Yin, J., Gordon, N., Naik, N., Clement, A. C., & Cane, M. A. (2002). Is the gulf stream responsible for Europe's mild winters? *Quarterly Journal of the Royal Meteorological Society, 128*(586), 2563–2586. https://doi.org/10.1256/qj.01.128.

Sheppard, J. J., Stratton, R. H., & Gazley, C. (1969). Pseudocolor as a means of image enhancement. *Optometry and Vision Science, 46*(10), 735–754.

Szejwach, G. (1982). Determination of semi-transparent cirrus cloud temperature from infrared radiances: Application to meteosat. *Journal of Applied Meteorology and Climatology, 21*(3), 384–393. https://doi.org/10.1175/1520-0450(1982)021<0384:DOSTCC>2.0.CO;2.

Walsh, J. E., & Chapman, W. L. (1998). Arctic cloud-radiation-temperature associations in observational data and atmospheric reanalyses. *Journal of Climate, 11*(11), 3030–3045. https://doi.org/10.1175/1520-0442(1998)011<3030:ACRTAI>2.0.CO;2.

Wilson, M. (2002). Six views of embodied cognition. *Psychonomic Bulletin & Review, 9*, 625–636. https://doi.org/10.3758/BF03196322.

Woods, T. L., Reed, S., Hsi, S., Woods, J. A., & Woods, M. R. (2016) Pilot study using the augmented reality sandbox to teach topographic maps and surficial processes in introductory geology labs. *Journal of Geoscience Education, 64*(3), 199–214. https://doi.org/10.5408/15-135.1.

Xie, C. (2011). Visualizing chemistry with infrared imaging. *Journal of Chemical Education, 88*(7), 881–885. https://doi.org/10.1021/ed1009656.

Ynnerman, A., Löwgren, J., & Tibell, L. (2018). Exploranation: A new science communication paradigm. *IEEE Computer Graphics and Applications, 38*(3), 13–20. https://doi.org/10.1109/MCG.2018.032421649.

Chapter 13
Conclusions and Future Outlook

Jesper Haglund, Fredrik Jeppsson, and Konrad J. Schönborn

The aim of this book project was to bring together international educators, researchers and practitioners on the use of Thermal Cameras in Science Education. The outcome was a collection of chapters describing the physics behind thermal cameras, empirical studies on their use in educational settings, and examples of their practical use for teaching.

In closing the volume, we first draw conclusions from the chapters by responding to the overarching question: *What are the current opportunities and challenges in introducing Thermal Cameras in Science Education, and what does the future hold?* The future outlook section emphasises potential next steps of research on the educational use of thermal cameras and how the efforts collated in this volume might impact teaching practice.

13.1 Conclusions from the Chapters of the Book

As the chapters demonstrate collectively, the range of topics in science education that may benefit from the introduction of thermal cameras is wide. Many of the studies described in the volume focus on thermal phenomena in physics education,

J. Haglund (✉)
Department of Engineering and Physics, Karlstad University, Karlstad, Sweden
e-mail: jesper.haglund@kau.se

F. Jeppsson
Department of Behavioural Sciences and Learning, Linköping University, Norrköping, Sweden
e-mail: fredrik.jeppsson@liu.se

K. J. Schönborn
Department of Science and Technology, Linköping University, Norrköping, Sweden
e-mail: konrad.schonborn@liu.se

J. Haglund et al. (eds.), *Thermal Cameras in Science Education*, Innovations in Science Education and Technology 26, https://doi.org/10.1007/978-3-030-85288-7_13

which may be the topic that springs to mind when a teacher first encounters this technology. For example, the empirical studies conducted in a school setting by Loukomies et al. and Åhman and Jeppsson focus on how heat conduction through solids can be made visible with infrared (IR) cameras.

In his chapter on the fundamentals of the physics and technology of thermal imaging, Vollmer describes how the technology relies on detection of IR radiation. Many of the chapters bring up how students can benefit from the use of IR cameras in studying thermal radiation as a phenomenon, and compare it with the properties of visible light. For example, Planinšič et al. describe how students can study how IR camera temperature measurements of surfaces depend on their reflective property. Similarly, Wong and Subramaniam show how Perspex is transparent in visible light, but opaque in the IR range.

Within the subject of physics, the potential of thermal imaging goes far beyond what is traditionally taught as thermal science. Many of the contributions, including those by Nordine et al., Planinšič et al., and Wong and Subramaniam point to the use of IR cameras in showing a temperature increase due to friction. Friction is typically used as an explanation of energy loss in mechanics, which can now be communicated visually. Within mechanics, Planinšič et al. also highlight the opportunity of quantitative modelling of energy conversion as a marble collides with and penetrates cardboard by measuring the temperature increase of the cardboard with an IR camera. In this regard, Nordine et al. demonstrate a quantitative positive effect of introducing IR cameras on students' learning of energy conservation and dissipation. Furthermore, Wong and Subramaniam show how IR cameras can be used to study the temperature increase of electrical components and thereby power dissipation, while Planinšič et al. apply the technology in teaching optics, by showing how IR radiation can be focused through reflection on a concave mirror.

Going beyond physics teaching, Wong and Subramaniam describe multiple applications of the technology for chemistry and biology teaching. For example, they visualize the temperature increase or decrease of different chemical reactions, including those of commercial heat packs and cold packs. Wong and Subramaniam, and Haglund et al. also show how IR cameras can be used to study animal temperature and thermal regulation. In addition, there are various other potential applications that have not been expanded upon in the book. For instance, in his literature review Samuelsson identifies a lack of studies in technology education.

As in the case of the range of topics, the chapters represent many different approaches to how IR cameras can be recruited in science teaching practice. Planinšič et al. suggest an epistemological framework for informing different uses of modern devices, such as IR cameras, in teaching. At the most basic level, students can use the device as a black box, more or less using the IR camera as a thermometer. At the next level, students can be encouraged to learn how the device works, for example, how IR cameras detect thermal radiation. At the third level, students are invited to learn new physics concepts by applying their knowledge of how the device works, as in the case of formulating and testing possible explanations for why containers of different material have very different temperature readings with IR cameras when boiling water is poured into them. In a similar fashion, Haglund et al. focus on how students can be encouraged to use IR cameras in open-ended inquiry activities, where they formulate questions and design experiments in relation to animal thermoregulation.

Although most of the chapters present IR cameras as a way to visualize thermal phenomena qualitatively by showing what surfaces are warmer or colder than others, or relative changes in temperature, there are also examples of using the technology for quantitative measurement. In particular, Oss provides examples of how IR camera measurements can be used in combination with theoretical modelling of heat transfer phenomena in university physics teaching. In addition, as noted previously, Planinšič et al. have engaged students in modelling energy conversion in collisions, with discussions of possible error sources in their calculations. However, as Vollmer argues in his chapter on the use of IR camera smartphone accessories, quantitative measurements with hand-held low-end IR cameras should be used with caution. In fact, due to measurement errors and the difficulty in interpreting the rendered IR images, he recommends using the technology primarily to visualize thermal phenomena in a qualitative way.

The types of low-end handheld IR cameras that Vollmer describe are those adopted and used in most of the studies described in the book. This is largely due to their lower prices and ease of use for teachers and students. The study by Lundin Palmerius and Schönborn is an exception, however, where the design of a more advanced augmented reality system is described for implementation as a science center exhibit, where IR camera imagery is projected onto the surfaces of the objects themselves.

Different theoretical perspectives are adopted as part of the empirical studies constituting Part II. Samuelsson's literature review of empirical studies on students' use of IR cameras in science education conveys that research in this area often applies a social semiotics perspective in analysing students' meaning-making processes when using IR cameras. Theories from social semiotics are also adopted in the chapter by Åhman and Jeppsson, in their analysis of the role of the teacher and pupils and the affordances of the IR camera in pupils' investigation of heat conduction. Similarly, Haglund et al. focus their analysis on how well students' dialogue reflects inquiry processes. Nordine et al. have explored students' conceptual understanding, including assessing their learning of energy conservation and dissipation. In contrast, Loukomies et al. focus their analysis on how IR cameras can stimulate positive emotions toward science among students, as expressed in students' video blogs when conducting science activities. These works provide more solid evidence in supporting the impression that many of us have had in that students find using IR cameras highly engaging and enjoyable, and that their learning can be stimulated by the technology since it allows them to interrogate otherwise invisible phenomena.

Although most studies have focused on opportunities for introducing IR cameras in science teaching, educational role players also have to recognise identified challenges with the technology. First, although prices have fallen in recent decades, acquiring a set of handheld IR cameras still represents a rather large cost for a typical school context. Convincing teachers that it is a worthwhile pedagogical investment requires that they are aware of how the technology can be used in teaching to optimally benefit students' learning and interest.

Second, although interpreting IR camera images is highly intuitive, it may lead to making erroneous conclusions about the temperature of surfaces. For example, as discussed by Vollmer and Planinšič et al. what appears as blue on the IR camera display does not always necessarily mean a lower temperature than red appearing objects, due to differences in emissivity. Samuelsson reviews research that has concluded that younger students benefit from a simple model of heat transfer to be able to interpret IR images of heat conduction through metals. Nordine et al. describe that one approach to countering difficulties of interpreting IR camera images is to begin a teaching sequence by having students take a "driver's licence" for using IR cameras, where they are asked to interpret and discuss challenging IR images. As emphasised by Vollmer, however, more advanced analysis requires formal training in thermography, and measurement errors and difficulties in interpreting thermal images limit the possibilities of using the technology for quantitative measurements in teaching.

Third, we would like to note some limitations of what IR cameras can bring to science teaching. Although IR camera imaging provides a dynamic, vivid and holistic view of macroscopic aspects of thermal phenomena, it does not directly support students' understanding of microscopic aspects, such as free electron models of heat conduction in metals. In addition, as low-end IR cameras are designed to measure surface temperatures of solids and liquids, traditional thermometers still play an important role in scientific inquiry of thermal phenomena, such as in the temperature measurement of gases and the bulk of liquids.

13.2 Thermal Cameras in Science Education: Future Outlook

As recognised by Charles Xie in the foreword, computers and computer-assisted programs have been given increasing attention in schools since the latter part of the twentieth century. One such example is the graphing calculator and another more recent example is the use of personal mobile phone technology in science classrooms as part of hands-on experiments (Wu et al., 2012). Several studies in this book, for example Åhman and Jeppsson, have shown how IR sensors can be connected to smartphones in school laboratories. In addition to concrete devices such as smartphones for plotting and visualising data, teachers and students also have access to demonstrations, open source simulation and visualization software on the internet. For instance, Xie (2014) has posted various video clip demonstrations of IR camera experiments. Furthermore, students and teachers can also utilise the internet as a way to conduct remote laboratory work (Lowe et al., 2013). One clear advantage for schools to engage remote laboratory work is that it is often free of charge and can easily be conducted online with a computer in a school classroom, while also being authentic.

As it is advantageous with, for example, free remote laboratory work, it is at the same time a challenge for schools to acquire funding to develop laboratory environments that require purchasing new laboratory equipment, such as investing in several thermal cameras. For IR cameras to become a part of the "standard" equipment of school science labs, in strong competition with other digital tools, IR camera manufacturers need to prioritise lower prices, robustness and minimum need for maintenance, such as software updates.

So far, as shown in Samuelsson's literature review and in most contributions of this book, our efforts have focused on providing examples of how IR cameras can be used in science teaching and in qualitative analyses of how small groups of students interact with the technology. As represented in the direction described by Nordine et al. and Loukomies et al. there is a need for more studies with larger numbers of students, and quantitative analysis of the effect of the technology on students' learning and affective responses. In addition, as described by Lundin Palmerius and Schönborn, there is a case for further research on how thermal visualization environments can be used to evoke engagement and communicate scientific knowledge to the general public in the context of science centers and museums. Lastly, as recognised by Samuelsson, the study by Goodhew et al. (2015) on how being shown thermal images of their houses has an impact on householders' energy use a year later may serve as further inspiration in this regard.

Another possible means to make IR cameras more attractive for teaching practice is to increase empirical research, beyond physics and thermal science. As indicated in this book, also demonstrating the affordances of IR cameras across wider contexts in biology, chemistry and physics and between multiple levels of education, makes the research results accessible to communities beyond science education alone. Doing so requires that we further develop our communication channels to a wider audience in order to reach out and demonstrate the educational power of thermal imaging. Another way forward is to increase interest in using IR cameras in school laboratory experiments by promoting practice-based research programs, where practice is seen as a natural collaborator with reseachers in design, implementation and evaluation. In this way, it may be easier for school decision and policy makers to motivate the procurement of thermal cameras for the classroom.

In closing, we hope that this book will serve as a meaningful resource for science educators and science education researchers in multiple international contexts on the use of thermal cameras across the domains of physics, chemistry and biology. We also hope that the volume will inspire new pedagogical opportunities and future research directions for exploring the role of thermal cameras in learning and teaching science.

References

Goodhew, J., Pahl, S., Auburn, T., & Goodhew, S. (2015). Making heat visible: Promoting energy conservation behaviors through thermal imaging. *Environment and Behavior, 47*(10), 1059–1088. https://doi.org/10.1177/0013916514546218.

Lowe, D., Newcombe, P., & Stumpers, B. (2013). Evaluation of the use of remote laboratories for secondary school science education. *Research in Science Education, 43*(3), 1197–1219.

Wu, W. H., Wu, Y. C. J., Chen, C. Y., Kao, H. Y., Lin, C. H., & Huang, S. H. (2012). Review of trends from mobile learning studies: A meta-analysis. *Computers & Education, 59*(2), 817–827.

Xie, C. (2014). *The Concord Consortium*. Infrared Tube. Retrieved from http://energy.concord.org/ir

The manufacturer's authorised representative in the EU is Springer Nature Customer Service Centre GmbH, Europaplatz 3, 69115 Heidelberg, Germany. If you have any concerns regarding our products, please contact ProductSafety@springernature.com

Printed and bound by CPI Group (UK) Ltd, Croydon, CR0 4YY

30/09/2026

02225491-0002